Gunilla Jörnson

Emin Tengström

Urban Transport Development

Gunilla Jönson
Emin Tengström

Urban Transport Development

A Complex Issue

With 33 Figures

Professor Gunilla Jönson
Professor Emin Tengström
LTHs kansli
Sölvegatan 22E
221 00 Lund
Sweden

Library of Congress Contro Number: 2005922817

ISBN-10 3-540-25376-9 Springer Berlin Heidleberg New York
ISBN-13 978-3-540-25376-1 Springer Berlin Heidelberg New York

Springer is a part of Springer Science+Business Media
springeronline.com

Printed in Germany

Typesetting: Data conversion by the author.
Final processing by PTP-Berlin Protago-TEX-Production GmbH, Germany
Cover-Design: medionet AG, Berlin
Printed on acid-free paper 62/3141/Yu – 5 4 3 2 1 0

Preface

In an age of rapid globalization, humankind is confronted with a number of challenges. One of these is the future development of urban transportation. More and more, big cities are plagued by growing transport problems and the familiar consequences of these problems.

One of the original ideas behind the founding of cities was to facilitate the number of contacts between various decision-makers, between sellers and buyers, between teachers and students for example, without increasing the need for mobility. In modern cities, particularly the big ones, the inhabitants have to move around a good deal to carry out their ordinary tasks and pursue their leisure activities. At the same time the volume of transported goods has grown enormously within cities as well as to and from cities. The historical change in urban transport can be summarized in three concepts: increasing urbanisation (associated with the growth of the world population), increasing motorisation and increasing car dependence.

Mobility of goods and people is today central to economic growth and social development. If big cities are to remain efficient and attractive places for development, the problems of urban transport have to be dealt with in one way or another. The issue of positive development of urban transportation seems, however, to be confused. This vague statement can be reformulated by the pertinent question: How does one cope with the complexity of urban transport development?

Around the beginning of the new millennium the board of The Volvo Research and Educational Foundations (VREF) made the decision to initiate a new programme. The objective of this programme is to inspire and support new and equitable solutions for urban transport, to improve accessibility and safety as well as efficiency and

environmental sustainability. (In the website www.volvoresearch foundations.com there is a link to the VREF Future Urban Transport Policy Statement Document.)
The aim of the VREF programme is to initiate and financially support:

- New scientific approaches for the understanding of how successful development may be planned and implemented in major urban areas
- Constructive processes of change though the combination of new knowledge with coordinated actions

The programme is founded on three cornerstones:

1. A number of globally distributed Centres of Excellence with the objective of establishing an international network of multidisciplinary and interdisciplinary co-operation through funding research projects of the highest international standard
2. A somewhat larger number of Smaller Projects, with the objective of supporting research and complementing the Centres of Excellence through funding smaller international research projects relevant to the main theme of the entire programme, emphasising the role of younger scientists in the development of new competence in this field
3. Regular international Future Urban Transport conferences, to create a meeting-place (and thereby increased interaction) for politicians, planners, industrialists, and researchers (especially the grant-holders of the programme)

A first conference on Future Urban Transport took place in Göteborg, Sweden in the year 2000. A second conference was organised in the same city in 2003. Its title was "Future Urban Transport: how to cope with the complexity of urban transport development".

This anthology is based on the second conference. The book should, however, not be seen as the proceedings of the conference. Instead, a number of presentations made at the conference have been selected and organised within a coherent structure consisting of five sections. We believe that many politicians, policy-makers, planners, consultants, industrialists, representatives of NGO's and researchers etc will find a number of interesting chapters in this anthology. Each author is, of course, responsible for the views expressed in his/her own contribution to this book.

The anthology will also hopefully inspire many of these groups of readers to participate in the next conference and to join our discussions. The next conference will take place in Göteborg in April 2006. For more information about this conference see www.fut.se

Göteborg, in November 2004

Arne Wittlöv

Chairman of the Board of the Volvo Research and Educational Foundations (VREF)

Acknowledgements

Volvo Research Foundations has initiated a programme (see the Preface) to inspire and finance new equitable solutions for urban transport through new scientific approaches for the understanding of how successful development may be created and implemented in larger urban areas and aid constructive processes of change. The programme is built on three corner stones 1/ a number of globally distributed Centres of Excellence 2/ a somewhat larger number of Smaller Projects, with the objective to support research, complementing the Centres of Excellence, and 3/ recurring international Future Urban Transport conferences, to create a meeting-place for politicians, planners, industrialists, and researchers. So far two conferences have taken place in 2000 and 2003.

This book presents a selection of presentations at the conference in 2003 in Göteborg, Sweden. The two editors wish to thank the many people, who have contributed their time and expertise to the conference. We thank the authors of the chapters of this book. They have not only benevolently contributed to the creation of the book but also and cooperated with the editors to finalize the manuscript. In addition, the editors want to acknowledge the work that Gunilla Bökmark has done as an assistant in preparation and the execution of the conference as well as in participation in the gathering of the contributions to the book. She has also transformed the text to meet the requirements of the publisher. The editors of this book also want to acknowledge the conference planning group including Rune Landin from the Volvo Group, Stephen Wallman from Volvo Cars and Peter Thormählen from the Volvo Research and Educational Foundations that aided in selection and structuring of the 2003 conference programme. Kikki Hugestrand is also acknowledged for her handling of the practical arrangements that made it possible for politicians, planners, industrialists, and researchers to meet for the constructive dialogue that lies behind the papers presented in this book.

The editors will also acknowledge and thank the students from different universities taking notes during the conference, to ensure that it was possible to draw conclusions that will be used to move the development work forward in the urban transport area based on the Volvo Research and Educational Foundations vision.

Finally, but not least, we wish to thank the board of Volvo Research and Educational Foundations that gave us the resources to make this book.

Göteborg, January 2005

Gunilla Jönson
Emin Tengström

Notes on the Contributors

Mr. Stephanos Anastasiadis

Mr. Stephanos Anastasiadis holds degrees in Psychology and Science of Religion from the University of Cape Town, and in international politics from Trinity College, Dublin; he was until recently Policy Officer at the European Federation for Transport and Environment (T&E), Europe's principal environmental organisation campaigning on transport issues, where he worked for five-and-a-half years; his main areas of responsibility were in the field of 'transport and society', which includes such topics as health and quality of life, urban transport, mobility and social change; his other areas of work at T&E included environmental integration, and governance; he left T&E in late 2004 to start an independent career.

Major Desmon Brown

Major Desmon Brown is a graduate of the Royal Military Academy Sandhurst and spent the first 15 years of his working life serving in the Jamaica Defense Force (JDF); after leaving the JDF, Major Brown started an engineering consultancy firm and a vehicle maintenance workshop; in 1995 he was selected to prepare new bus specifications for city buses for the new public transportation system in the Kingston Metropolitan Transport Region in Jamaica; in 1996 he was employed as the Operations Manager of the holding company established for the public transport assets to be provided by the government - Metropolitan Management Transport Holding Limited; in 1999 he was promoted to the position of General Manager; Major Brown holds a Masters Degree in Public Transport Management and is a chartered member of the Institute of Logistic & Transport.

Dr. Annika Carlsson Kanyama

Dr. Annika Carlsson Kanyama holds a Ph.D. in Natural Resource Management from Stockholm University and is currently working as a research leader in Environmental Strategies Research Group at The Swedish Defence Research Agency in Stockholm; she has specialised in issues related to sustainable consumption including food, transport and housing; she has substantial experience of working in interdisciplinary and international settings including experiences from Tanzania and South America.

Professor Robert Cervero

Professor Robert Cervero is active at the Department of City and Regional Planning, University of California, Berkeley; he earned his Ph.D. in Urban Planning, University of California, Los Angeles in 1980; Professor Cervero is the author of a variety of books, research reports and articles as well as being an invited consultant, speaker and also visiting professor in various organizations and countries.

Mr. Andrew Cotugno

Mr. Andrew Cotugno received a Bachelor's degree in city and regional planning from California Polytechnic State University in 1974, and has done graduate work in public administration at Lewis and Clark College in Portland; he has more than 25 years of professional experience in the transportation and planning fields; prior to his current position he worked as a transportation planner for both Metro and the Mid-Ohio Regional Planning Commission; Mr. Cotugno was appointed as Metro's director of Transportation in 1980; he is chair of Metro's Transportation Policy Alternatives Committee and Metropolitan Technical Advisory Committee and is active in regional, state and federal financing activities for transportation and growth management projects.

Dr. Graham Crampton

Dr. Graham Crampton is a Lecturer in Economics at Reading University, UK; he holds an Economics Ph.D. from Cambridge, England; in his research, he specializes in urban transport and other areas in urban and regional economics; recently, his work has had an international comparative emphasis; he has contributed to numerous research projects carried out by Environmental and Transport Planning, focusing principally on work on economics, statistics, GIS spatial analysis, and British policy-related interviews.

Professor Carmen Hass-Klau

Professor Carmen Hass-Klau (Dipl Ing, Msc, Ph.D. FCIT) is Professor in Civil Engineering at the Department of Transport and Public Transport Systems in Europe at the University of Wuppertal, Germany; she is also Principal of Environmental and Transport Planning, which is a research organisation based in Brighton and specialising in the field of public transport, mainly light rail, pedestrianisation, the environment, cycling and traffic calming.

Professor Gunilla Jönson

Professor Gunilla Jönson is Professor of packaging logistics at Lund University, Sweden and Adjunct Professor at Michigan State University, USA; at present she is also Dean of the faculty of technology at Lund University and sits on several Swedish academic and industrial boards; she has extensive experience of different packaging logistics issues from her time at SCA Packaging in Sweden and of their operations in Europe, especially in Belgium and the UK; she has worked on packaging related questions with UN/ITC and USAID in the People's Republic of China, Thailand and South and Central America; she has now around 10 Ph.D. students doing research in the area of consumer demands on packaging and in the area of product development and innovation to reach efficient, effective, safe and sustainable distribution, especially food.

Professor Arne Kaijser

Professor Arne Kaijser is Professor of History of Technology and head of the Department of History of Science and Technology at the Royal Institute of Technology in Stockholm, Sweden; he holds a M.Sc. in Technical Physics (1973), and a Ph.D. in Technology and Social Change (1986) from Linköping University; besides his academic career, he has worked for seven years as civil servant in various agencies with futures studies, research policy and development aid; his main research interest concerns the historical development of infrastructural systems; he has studied different kinds of such systems (energy, transportation, communication and water systems) in several countries (Sweden, the Nordic countries, the Netherlands and the US), and he is particularly interested in the dynamics of such systems, in the shaping of their institutional frameworks, and in the integration of systems across national borders.

Dr. Ahmad Kanyama

Dr. Ahmad Kanyama took a Ph.D. in Build Environment Analysis at the Royal Institute of Technology in Stockholm, Sweden, in 1995; before embarking on research studies he worked as urban planner in the capital city of Tanzania for 7 years; his work since 1995 as a consultant and a researcher has been focused on urban planning and sustainable urban transportation; he is currently conducting research at the Royal Institute of Technology; Department of Infrastructure, Division of Urban Studies.

Professor Bengt Kasemo

Professor Bengt Kasemo is Chairman of the VREF Scientific Council; he is since 1983 Professor of physics at Chalmers University of Technology, Göteborg; he heads a research group of about 40 people active in basic surface science, biomaterials and biointerfaces, nanoscience and nanotechnology and heterogenous catalysis for environment and sustainable energy; he has served on national science policy and research funding boards, including the Swedish Government's Advisory Board for Research; he is a member of the Royal Swedish Academy of Sciences and of the Royal Swedish Academy of Engineering Sciences (previous vice president of the latter); he has received several prices e.g. the George Winter award from the European Society for Biomaterials (1999) and the Akzo-Nobel prize from the Royal Academy of Engineering sciences (2001).

Professor Anna-Lisa Lindén

Professor Anna-Lisa Lindén is Professor of Sociology at Lund University, Sweden; she has written several books and articles within her fields of research e.g. travel patterns and environmental impacts, the efficiency of policy instruments on individual behaviour, social differentiation in urban systems, lifestyles and public health; several of her projects are interdisciplinary, including researchers from economics, business administration, systems ecology, medicine and transportation; at the Department of Sociology Professor Lindén has the responsibility for the planning and administration of postgraduate education; she is board and/or committee member in national research funding agencies.

Mrs. Liya Liu

Mrs. Liya Liu is an expert in the field of comprehensive transportation development strategy studies at the Institute of Comprehensive Transportation of the National Development and Reform Commission (NDRC) in China; she has been involved in many domestic and international transportation study projects since 1986; she has managed "the Green Urban Transportation Program" in China sponsored by the former W. Alton Jones Foundation; she spearheaded several BRT development initiatives such as organizing and conducting BRT seminars to the directors at the NDRC, governors and mayors in several provinces and large cities in China; she recommended Kunming city to build a BRT network and this suggestion was adopted in the Kunming Urban Transport Development Plan in 2002; Mrs. Liya Liu has also submitted a BRT development policy suggestion to Beijing's mayor; since 2003 Mrs. Liya Liu has initiated the "Urban Transport Development Strategy Partnership Demonstration Program in China".

Dr. John Lupala

Dr. John Lupala took a Ph.D. in Build Environment Analysis at the Royal Institute of Technology, Stockholm, in 2002; he is now a Lecturer and Head of the Department of Architecture of the University

College of Lands and Architectural Studies (UCLAS), a constituent College of the University of Dar-es-Salaam, Tanzania; before joining UCLAS in 1991, Dr. Lupala worked as an Urban Planner in the Capital Development Authority in the capital city of Tanzania from 1987 to 1991; in his work since 2002 as a Lecturer and researcher he has focused on urban planning and sustainable urban transportation, participatory approaches in settlement planning and urban housing focusing on cooperative societies and settlement upgrading.

Managing Director Måns Lönnroth

Managing Director Måns Lönnroth has been active in many administrative and political areas after graduate studies in applied mathematics at the Royal Institute of Technology in Stockholm, Sweden; he was a member of the Delegation for Energy Policy, Prime Minister's Office 1974-1975; a member of the Secretariat for Futures Studies 1975-1984; he was appointed Assistant Professor (Docent) at the Department for technology and social change, University of Linköping in 1980; he was member of Stockholm County Council 1979-1994; political advisor in the Prime Minister's office 1984-1991; state secretary at the ministry of Environment 1994-1999 and he is now (since 2000) Managing Director of MISTRA (Foundation for Strategic Environment Research).

Professor David Marks

Professor David H. Marks received his Ph.D. in Environmental Engineering (1969) from The Johns Hopkins University; he joined the Department of Civil Engineering at MIT in 1969; from 1985 until 1992 he served as the Head of the Department of Civil Engineering; in 1993 he was awarded the James Mason Crafts Professorship endowed Chair; in 1999, he was awarded the Morton (1942) and Claire Goulder and Family Endowed Chair in Environmental Systems; in May 2002, Chalmers University of Technology in Sweden conferred on him the degree of Honorary Doctor; he is the founding director of the MIT Laboratory for Energy and the Environment and the MIT coordinator for the alliance for global sustainability with MIT,

Chalmers University of Technology (Sweden), the Swiss Federal Institute of Technology (Zürich) and the University of Tokyo, Japan.

Dr. Paul Mees

Dr. Paul Mees teaches transport and land use planning, metropolitan strategic planning and planning law at the University of Melbourne since1998; before becoming an academic he was a lawyer; he was President of the Public Transport Users Association (Vic) from 1992 to 2001; he is the author of *A Very Public Solution: Transport in the Dispersed City* (2000); he has been a consultant to local, regional and State government transport agencies in Australia and New Zealand.

Professor Dinesh Mohan

Professor Dinesh Mohan received a Bachelor's degree in mechanical engineering from the Indian Institute of Technology (Bombay), a Master's degree in mechanical and aerospace engineering from the University of Delaware (USA), and a Master's degree and Ph.D. in Bioengineering from the University of Michigan (USA); he is now Henry Ford Professor for Biomechanics and Transportation Safety, and Co-ordinator of the Transportation Research and Injury Prevention Programme at the Indian Institute of Technology, Delhi; amongst other posts Professor Mohan has been a consultant on safety related matters to government departments in various countries, automotive industries and also to international organisations like the World Bank and WHO.

Professor Wilson Odero

Professor Wilson Odero is a medical doctor and public health practitioner; he holds an M.Sc. degree and Ph.D. in Epidemiology and Public Health from the London School of Hygiene & Tropical Medicine; he is presently an Associate Professor and Dean of Moi University's School of Public Health in Kenya; Professor Odero has published widely on injury epidemiology and control, and contributed to the understanding of alcohol as a risk factor in road traffic injuries in developing countries; his current research includes establishing computerized medical records for injury surveillance in a health centre, and conducting a case-control study on causes of road traffic crashes in an urban environment in Kenya.

Mr. Hans Örn

Mr. Hans Örn holds an M.Sc. in civil engineering from the Chalmers University of Technology in Göteborg, Sweden; with experience from both the public sector (as a transport planner in Göteborg) and the private sector (the Swedish vehicle industry), he has also spent some 15 years as a private consultant in urban transportation; he has lived in Peru and Singapore and his working languages are Swedish, English, Spanish and German; his present activities focus on developing countries in Asia, Africa and Latin America, often in projects supported by international organizations.

Professor Ethan Seltzer

Professor Ethan Seltzer majored in Biology and received his Bachelors degree with Distinction from Swarthmore College, in Swarthmore, Pennsylvania, in 1976; he received a Master of Regional Planning degree in 1979 and a Doctorate in City and Regional Planning in 1983 from the University of Pennsylvania; Professor Seltzer is currently serving as the Director of the School of Urban Studies and Planning at Portland State University in Portland, Oregon; from September 1992 through August 2003 he was the founding director of Portland State's Institute of Portland Metropolitan Studies; prior to joining Portland State University, he held positions as the Land Use Supervisor for Metro, the regional government in the Portland area, and as the Assistant Coordinator for the Southeast Uplift Neighbourhood Program.

Professor Emin Tengström

Professor Emin Tengström received a Ph.D. in Latin in 1964; he was Assistant Professor (Docent) in Latin at Göteborg University, Sweden from 1971; he was elected Chairman of the board of "The Centre for Interdisciplinary Studies of the Human Condition" in 1972;

he was Professor of Human Ecology at the Technical University of Chalmers, Göteborg in 1981; and Professor of Human Ecology at Göteborg University in 1986; he was Research Professor of Transportation Studies at Aalborg University, Denmark 1995 – 2001; he was member of the Scientific Council of VREF 1994-2003.

Dr. Marie Thynell

Dr. Marie Thynell holds an L.SSc and a Ph.D. in Peace and Development Research from Göteborg University, Sweden; since 1995 Dr. Thynell is a researcher, teacher and supervisor active in the fields of International Relations, Development and International Cooperation as well as in the fields of Science and Technology Studies and Latin American Studies; her present research focus is on transport policies, urban change, political order and societal development; she addresses particularly the social science aspects of international structures and political responses to urban transport demand, environmental problems and social inclusion/exclusion. She has made field studies on these topics in a number of cities, e.g. in Tehran, New Delhi, Dhaka, Rome, Copenhagen, Warsaw, Santiago and Brasilia.

Dr. Geetam Tiwari

Dr. Geetam Tiwari's educational background includes a Ph.D. from the University of Illinois, Chicago, Transport Planning and Policy; she has 15 years of professional experience in the areas of Transport Planning, Traffic Safety and Traffic Engineering in India, USA and Bangladesh; Dr. Tiwari has taught courses in International Traffic Safety training programmes in Australia, South Africa, Uganda, the Netherlands and India; her area of specialisation includes urban transportation planning including analytical and simulation models of environment, energy and traffic safety; her research has focussed on demand and infrastructure required for non-motorized transport and bus systems in cities and intercity highways and traffic safety issues in mixed traffic.

Professor John Whitelegg

Professor John Whitelegg started his career as an economic development; since then he has worked mainly at Lancaster University but also for three years as an official in the Department of Transport in Düsseldorf, Germany, and in India where he made a special study of non-motorised transport; Professor Whitelegg is the Managing Director of Eco-Logica Ltd, a transport consultancy based in Lancaster, as well as being Professor of Sustainable Transport at Liverpool John Moores University, Professor in the Department of Biology at York University and Leader of the Implementing Sustainability Group at the Stockholm Environment Institute, University of York; he is an invited member of the European Commission expert working group on sustainable urban transport; he is also a member of the UITP (international public transport organisation) Council on Sustainable Development.

Dr. Arne Wittlöv

Dr. Arne Wittlöv, Chairman of the Volvo Foundations for Research and Education, has a M. Sc. in Mechanical Engineering and is a Doctor of Technology *honoris causa*; he is Chairman of Göteborg University, member of the board of the Swedish Research Council, member of the Royal Swedish Academy of Engineering Sciences and member of the Royal Swedish Academy of Sciences; he is Chairman or board member of several Volvo companies and foundations; he has been President and CEO of Volvo Aero Corporation Executive and Vice President of the Volvo Group with main responsibility for technology – retired in 2001; Chairman of the Royal Swedish Academy of Engineering Sciences 2002-2004.

Contents

Section one: Some Basic Perspectives

Section two: Urban Transport Systems Today: a Variety of Complexity

Section five: Challenges in Dealing with the Complexity of Urban Transport

Section one: Some Basic Perspectives

Introduction

Reading this anthology is a way of getting involved in the process initiated by Volvo Research and Educational Foundations (VREF). The book is organised in five sections, all of them provided with an introduction that can be seen as an executive summary of the chapters belonging to each section. Through these introductions the reader is informed of the content of the section and is thus able to select certain chapters for close reading.

In the first chapter, Bengt Kasemo, chairman of the Scientific Council of VREF, delineates – in very general terms - *the new research area* that VREF is initiating and supporting (according to the policy programme described by Arne Wittlöv in the preface). Bengt Kasemo emphasizes that the programme should be truly international and not restricted to the so-called developed world but address urban transport development in less motorized countries as well.

Kasemo elaborates particularly on some aspects of this programme. A new understanding of urban transport development necessitates a more or less *holistic view* of the urban transport system. Such a view can only be realised by means of some kind of *system approach*. An urban transport system must be seen as a *complex system*, the long-term changes of which are difficult to predict. Even if the driving forces are fairly well known, the consequences of decisions taken may be unplanned or sometimes unexpected.

The most interesting aspect of an urban transport system is, therefore, not only to analyse *its structure and present function* but also to focus on the *dynamics of the development of the system*. This development is fairly easy to describe and analyse in the short term, but the most interesting thing is to know more about the dynamics of long-term changes. Here questions arise that are far from easy to answer. What actors initiate such changes? Are there circumstances that delay or even impede changes desired by a majority of the population of the city?

Secondly, Kasemo draws attention to the necessity and value of trying to distinguish between what is generic and what is specific in the development of each individual city's transport system. The generic content of a system is those components that are contained in essentially every system. The remaining components that are more or less specific vary from system to system. If one can identify the generic content, this forms a solid base for analyzing and comparing different systems. If the generic content is very small and the specific components totally dominate, the transferability of any programme/solution for transport development may be small, and vice versa.

Kasemo also emphasizes the question of energy and environment in urban transport development. He points to the "rapidly increasing awareness that fossil fuels are finite; oil and natural gas production worldwide are predicted to peak within one or a few decades, while coal has a significantly longer time scale – but still finite". He adds that the energy supply should be clean enough not to cause damage to health and the local environment or contribute to global warming. This makes the challenge of urban transport development even greater.

In the second chapter, Arne Kaijser introduces an elaborated theoretical perspective on urban transport systems and their development. He takes as his starting-point the theory of technology known as the theory of Large Technical Systems (briefly called the "LTS-approach"). It is mainly used for studies of electrical systems or telecommunications systems etc. Applied to the world of urban transport systems, the theory has to be adapted to the fact that an urban transport system is to be seen as a *megasystem* consisting of several *subsystems* which – in their turn – may be categorized in two groups: private transport and public transport.

Such a large technical system is, however, rather to be described as a *socio-technical system.* It includes *technical components* such as networks of roads (streets) and railways, nodes such as parking lots and stations and a variety of vehicles. It also includes *social components* that consisting of actors, institutions (both material and immaterial) and organisations. Public transport is driven by various organisations, while private transport is mostly run by individuals who make their choices independently of each other.

By means of the networks and the vehicles, more or less intensive *traffic flows* are generated. For the survival of the city and its inhabitants there is an amount of traffic of people as well as of goods (for instance food and of garbage) moving within the system. These flows are moving within the city or moving into or out of the city. A salient characteristic of urban transport is that several subsystems use the same networks of streets. This fact often leads to conflicts between the users of different subsystems.

The most intricate questions of urban transport megasystems are associated with their *long-term dynamics*. Socio-technical systems tend to pass through *successive phases*, each characterized by critical problems. It is possible to distinguish between at least four such phases in the history of urban transport, at least in cities in Europe and North America.

The long-term dynamics are basically determined by an interplay between networks of various actors and actor groups. This interplay may be successful or not. In the first case, the result may be certain *city-regimes* strongly influencing the development of the urban transport system.

The success of this interplay is, however, often hampered by the fact that the system in itself has a certain *momentum* that makes it difficult to implement bigger changes. One, therefore, talks about the *path-dependency* of a system.

In the third chapter, Emin Tengström discusses the words 'complex' and 'complexity' in relation to urban transport. These words can, on the one hand, be used in ordinary language with a somewhat vague meaning. As scientific terms, on the other hand, the *connotations* of the words are reduced or eliminated and the *denotations* of the terms are strictly defined.

If one regards urban transport as a complex system, one may describe its *structure and function* in ordinary language analysing its *interlinked problems*. It is also possible to discuss the complexity of *the process of change* of urban transport development in non-technical language.

If one instead wants to analyse the structure of an urban transport system and the process of its changes over time by a *systems approach*, one needs the perspective introduced by Bengt Kasemo and a theory of the kind described by Arne Kaiser.

A developed systems approach seems to be a highly desirable knowledge base for the future, as the development of urban transport systems today seems to be exposed to *market failures* as well as to *government failures* and *interaction failures* (this term refers to inadequate interaction between various categories of actors and between the local municipality and the market). It might be useful to add the concept of '*public acceptance failure*' to refer to the unwillingness of the general public to accept various proposals for reducing the problems of an urban transport system.

In using the words 'complex' or 'complexity' in relation to urban transport development, one has always to be aware of the difference between their meanings when used as scientific terms and when used as a part of colloquial language.

Chapter 1. Towards a More Holistic Approach to Urban Transport Studies

Bengt Kasemo

When I was appointed chair of the Scientific Council (SC) of the Volvo Research and Educational Foundations (VREF), the VREF Board had just taken the decision to reorient the funding and other activities towards Future Urban Transport (FUT). The SC's immediate task, and mine, was to build a suitable structure of the program. Some important parameters were: (i) The program should be truly international and not restricted to the so-called developed world but also address issues related to FUT in less motorized countries. (ii) The program should approach the FUT area in a system-oriented way, primarily looking at the whole system rather than at individual components or subsystems, like technical solutions. The latter was explicitly underlined in the subtitle to the FUT program; "How to cope with complexity". (iii) The approach was, of course, partly analytical and knowledge-oriented, but equally important was the goal to promote identification and implementation of desirable FUT solutions. (iv) The projects should be of a high international standard, and the major projects should be launched only after a careful international peer review process.

As a result of the initial work of the SC and Board of VREF, we have today a FUT program that is built on three corner stones: Globally distributed Centres of Excellence (CoE), Smaller Projects (SP) and tri-annual conferences on Future Urban Transport. To date three CoEs have been launched, at New Delhi Institute of Technology (TRIPP), India, at Columbia University, NY, USA and at UC Berkeley, Cal., USA. Two FUT conferences have been held.

The aim of the recurring international FUT conferences is to create a meeting-place for greater interaction, exchange of ideas and experience between politicians, city planners, industrialists, and academics, including the research grant holders within the FUT programme. The two FUT conferences so far were held in March 2000 and in September 2003, both times in Göteborg, Sweden.

What have we learnt to date? One of the aims of the 2000 FUT conference was to get external input regarding the selection criteria, content and direction of CoEs and SPs, i.e. to incorporate the results and messages of the conference in VREF's process to reorient the funding activities towards the FUT area. The conference topics were: *1: History and geography of urban transport; perspectives on human behaviour and the role of infrastructure. 2: Function and efficiency of transport. 3: Energy consumption and environmental aspects. 4: Safety aspects - prevention of traffic accidents and the care of injured persons in large urban areas. 5: Critical issues for the future - needs and actions (the latter was an attempt to sum up the conference, and identify paths for the future).*

There was a strong positive response from the participants to the initiative to gather actors from different parts of society; researchers, city planners, politicians, transport system providers, etc. There was, furthermore, strong support for an integrated approach to the various components of urban transport, for example *analysis, knowledge generation, actors, implementation, process of change, etc.* A major theme of the future work of VREF could be "*how to cope with complexity*".

The second conference and its Academic Workshop brought up a number of new questions and produced new insight. Some of the most important ones were (some of them are also articulated below):

- What does "coping with complexity" mean? What is complexity?
- What is generic and what is specific (local, contextual) in urban transport solutions? This is connected with so-called "path dependencies" and the possibility/difficulty of translating a successful solution for one city to another (transferability)
- Need for a holistic view on FUT

- Inclusiveness – the importance of the people and involvement of the public
- The role of financial solutions (who will pay) and the role of institutions
- Implementation schemes
- Energy and environment issues

Holistic view and systemic aspects

Technical solutions to urban transport problems obviously constitute only one component in the total "tool box". The historical path on which the city has developed and is evolving, the decision makers (government etc), the local and global economy, the public and the economic life of the city, which are using and demanding transport solutions, all have their influences. Noting this, and after having listened to the speakers at the two FUT conferences and to Professor Emin Tengström's pledge for "coping with complexity", it is obvious that the technical solutions are the easy part of the total challenge, and that a holistic and truly multidisciplinary approach is called for, using and integrating all the relevant knowledge platforms from social sciences, liberal arts, natural sciences and engineering.

Comment on complex systems

The word complexity rings different bells for different persons, as articulated by Professor Emin Tengström (Chapter 3). As a physicist, my thoughts turn to systems that are strongly non-linear and far from equilibrium. These are systems that usually have many different stable states, and can switch between these states upon external perturbations. (In extreme cases, when the "jumping between states" occurs in a special way, we talk about chaotic systems. In contrast, under other conditions, one of these (meta)stable states may be the single state for a long time)

The most famous example of such systems is probably weather and climate. Who has not heard about the metaphor of a butterfly lifting from a flower in Brazil (or somewhere else), thereby initiating a turbulent structure, which amplifies and gives rise to a hurricane in Japan (or somewhere else) some weeks later? The metaphor illustrates two things; Firstly, some very small perturbation in a strongly nonlinear (complex) system may have an enormous leverage and amplify the seemingly negligible perturbation, to create a large macroscopic perturbation. Secondly, the example illustrates the lack of exact predictability of such systems. Even if everything were known with high precision at the time, the butterfly lifted from the flower, there is some remaining uncertainty, that could take the system along the hurricane "path", or along another path where the turbulent structure just died out.

To generalize; even if we know a lot about a particular complex system, including its past history (path) and its present status, and even if we have a truly holistic and systemic view of things, there is still always some remaining little uncertainty that can switch the system into a new unplanned (maybe also unexpected)[1] path. This is what the "butterfly in Brazil" is aimed at illustrating; one would regard it as a marginal, unimportant event for the weather, and in most cases it may have no effect. However, sometimes– maybe in combination with other seemingly marginal co-factors – it might change the path into a new one. This is something we have to live with - and actually do live with in our daily lives!

There is no reason to suppose that planning and implementation of FUT is not a complex system of this kind. Then it forces us to be continuously alert and prepared to take corrective actions, because the development may take new unexpected or unplanned[1] paths. With such preparedness there is no reason to be paralyzed by the "complexity" – it is just a matter of realizing (almost as a trivial conclusion) that corrections of prepared plans are needed. It merely illustrates the dynamics of the development of a complex system. It also illustrates that planning based on multiple scenarios is valuable. By analyzing the different starting conditions of a system (on what

[1] "Unplanned" does not necessarily mean unexpected. If one has planned well and is open-minded, by having formulated a number of scenarios about what might happen, the new path may not be a surprise, even if it is unplanned and therefore probably undesirable.

historical path has it developed? where are we just now?), one learns which factors in the system its development is particularly sensitive to, and which factors are not so influential.

This is a rational and constructive way of looking at complexity: if the system is complex in the above sense – and urban systems are certainly of this kind - one cannot escape from the complexity, but one can build into the system ways of reacting early to, and correcting, undesired paths, if and when they are encountered. The latter requires careful follow-up of the development, and insight into the fact that we are dealing with dynamic systems. The earlier one can discover undesired development and take corrective action against it, the simpler and less costly is the correction. An opposite and non-rational – at worst paralyzing - way of looking at complexity is to turn away from the concept and say "this is too complex – I cannot deal with it, or try to find linear solutions (bound to fail)". (Living organisms, like ourselves, are wonderful examples of strongly non-linear systems, which have developed a remarkable control and correction system that keeps the system - in most cases - on the right stable path).

What is generic and what is specific?

At the second conference this topic emerged (at least to me) as one of the key questions. One context when it arose was in connection with the question of transferability – why are good examples/solutions difficult to translate/implement at other places? Is every city so unique (specific) that transferability is hampered for that reason? Or are there other reasons?

The generic content of a system, e.g., an urban transport system, are those components that are "robust" in the sense that they are contained in essentially every system. The remaining components are more or less specific and vary from system to system. If one can identify the generic content, this forms a solid base for analyzing and comparing different systems. If the generic content is very small and the specific components totally dominate, both transferability between, and the opportunities to compare, different cities may be small and vice versa.

One can identify several trivial generic components of urban transport systems, like people and goods needing transport, the need for financial solutions, the need for energy to drive the transport systems, the need for organized institutions or companies to operate the systems etc. There are also deeper and less obvious aspects that are not so easily distinguished as generic or specific. Take, for example, the past history of a particular city. The historical development constitutes a "path" along which the city – and its transport system - has developed, and which sets parameters for the future path/development. Is this a specific component or a generic one? It is, for example, tempting to say that the history of Rome constitutes a specific component for Rome, and the history of New York a different one for New York, and that in this respect they lack this common (generic) component. However, the very fact that the historical paths are important for both cities and influence the opportunities to find future solutions is a generic component.

I do not intend – and I am not even able – to pursue a more detailed discussion on this topic. The point is that it might be helpful to have a more thorough discussion on generic versus specific aspects of transport systems and solutions. Such analysis and discussion might be rewarding with respect to both understanding how transport systems may be changed and to identifying the barriers to such change.

Energy and Environment

There is a rapidly increasing awareness that fossil fuels are finite; oil and natural gas production world-wide is predicted to peak within one or a few decades while coal has a significantly longer time scale – but are still a finite resource. Even if these estimates were wrong by factors of two or more – which seems unlikely – we are still facing the reality that fossil fuels are finite and that their prices will rise. Since the so-called provision of sustainable energy – essentially solar (wind, photovoltaics, biomass, wave, solar thermal) and geothermal – are developing at a slow pace, it seems almost inevitable that we (=world society) for quite a while (many decades) will have to rely on coal and nuclear power, in a transition period between the oil/gas society and a so called sustainable energy society.

Of course this position can and should be challenged. Maybe solar energy in its different forms will develop faster than I and many others believe. Maybe new yet unidentified fossil resources will be discovered. Maybe fusion technology will finally be mastered. However, these hopes are in my view too faint to build the future on.

To provide an energy supply where the availability and price per kWh develop along a non-disruptive path is in any case decisive for what we call a sustainable society, for its production of goods and services, housing, water and food, a good environment, etc., and thus also for sustainable urban transport systems. This constitutes a significant uncertainty factor for FUT systems, which will affect which paths that FUT systems develop along. A smooth and evolutionary transition may be possible (and is what we hope for), but a more disruptive scenario is not unlikely. We just don't know. This is also part of what we call path dependencies and of "coping with complexity".

Please note that the above was said without taking the environmental factors into account. Adding as an additional condition that the energy supply should preferably be clean enough to avoid environmental effects both on the local scale (e.g. health effects), and global scale (green house gases, global warming) makes the challenge even bigger.

The point here is not to advocate an "energy shortage scare", but to emphasize one more important factor that adds to the complexity of understanding, developing and planning for FUT systems.

Chapter 2. How to Describe Large Technical Systems and Their Changes over Time?

Arne Kaijser

A system of systems

Complexity is one of the characteristics of urban transport. But how can we describe this complexity? What models and concepts could be fruitful in trying to understand and deal with this complexity? In my chapter, I will apply concepts and perspectives from a theoretical point of view developed within my own field, the history of technology. It is called the Large Technical System approach, or the LTS approach. This approach was first developed by Thomas P. Hughes in the magnificent book *Networks of Power*, which analyses the development of electricity systems in the US, Britain and Germany.

The LTS approach has mainly been used for studying the development of electricity systems, telecommunication systems, railway systems and similar, rather well defined systems. The urban transport system is of a somewhat different kind. It can be described as a megasystem, *a system of systems*. But even if it is different I still think that some of the concepts and perspectives from the LTS approach can be fruitful in describing and analysing urban transport systems. And I further believe that comparisons between urban transport systems and other man-made complex systems can be a way to acquire new perspectives on urban transport systems. I should point out that I will primarily talk about cities in the Western world as I am not so familiar with cities in other regions of the world.

A basic point of departure in the LTS approach is to view systems not as purely technical systems but as socio-technical systems. This means that not only the technical components and installations are taken into account, but also the people and organizations that build, operate and use all the components, as well as the institutional framework in which they carry out their activities.

How then can one describe an urban transportation system as a socio-technical system? The *technical* components of the system consist of networks of railways and roads, of nodes such as stations and parking lots, and of vehicles of many different kinds. These components make up subsystems, which in turn are often grouped together into two main categories: public transport and private transport. Goods traffic using trucks is another important subsystem. And then one should not forget that the networks of roads and streets are also essential for people who are walking or cycling.

All these networks and vehicles provide two kinds of flows: of people and goods, and these two flows have quite different characters. People travel primarily *within* an urban region, commuting to their jobs or schools, making service trips, visiting friends etc. Only a rather small fraction of the people travelling in a city area are simply passing through, or coming to visit. It is quite different with goods. They flow primarily *into the city or out of it.* Every city is totally dependent on regular supplies of food, beverages and other necessities consumed by its inhabitants. And there are equally regular flows of garbage in the other direction.

Let us now turn to the organisational and managerial aspects of the different subsystems, which are characterised by startling contrasts. On the one extreme we have the large number of individual car-owners who can use their car more or less as they like and who are personally operating their vehicle. There are various ways that have been developed to try and influence their behaviour, including traffic rules and signs, fuel taxes and parking fees. But basically car traffic is almost anarchistic in character, and each car driver is king (more often then queen) in his car. However, sitting in queues each day, he hardly feels like he is king of the road. On the other extreme we have public transport. Here the travellers are passengers in large vehicles operated by professional drivers, which are in turn employed by large organisations. The organisational complexity of public transport is very high, particularly in cities with many transport modes operating across boundaries of autonomous municipalities.

What I see as a salient characteristic of urban transport compared with many other infrastructure systems, is that many subsystems share the same network, namely the roads and streets. When traffic increases and road space becomes scarce, this leads to fierce competition. As many of you know, tram systems were phased out in many European and American cities in the mid-20th century. This was not primarily because they were technically or economically inferior. No, they were squeezed out from the streets by systems with more powerful backing. Here we have an important topic for research in my opinion: how have conceptions of roads and streets changed over time and how have the rules for access to streets and roads been altered.

This leads me on to a related issue. In the book *Networks of Power* Thomas P Hughes shows that managers of electricity systems a hundred years ago were almost obsessed by what they called the "load factor". In the early days of electricity most of the power was used for lighting, and this meant that the full generating capacity was only used a few hours each day. Samuel Insull, who was manager of the Chicago power company, was the first to realise fully that this was very uneconomical, and he found solutions to improve what he called the load factor, i.e. the rate of the power actually generated in relation to the potential maximum production. He started to sell electricity at a lower price to industrial consumers, which mainly needed power during the daytime. But he kept high prices for ordinary customers who mainly used power at peak periods. What Insull introduced could be called congestion pricing, and he managed to obtain a much better load factor.

Managers of other infrastructural systems have learnt about the importance of the Load factor, and have introduced similar pricing policies. We are all used to making mobile phone calls in the evenings if possible, and we adjust our trips by air to APEX conditions. For me it is difficult to understand why the Load factor concept has been so absent in the urban transport domain, and why the resistance to congestion charges has been so fierce.

Long-term dynamics

So far I have mainly discussed the daily operations of urban transport systems, what could be called their short-term dynamics. But these systems also have long-term dynamics in close interaction with what could be called the urban landscape with all its buildings and other structures. The building of roads and railways influence where new dwellings, factories and offices are built, and when people start to live or work in these new buildings this influences the transportation demands, calling for additional vehicles, new roads and railways etc. In industrialised countries there has been a growing ambition over time to co-ordinate such processes through elaborated plans, but often the actual development, at least partly, departs from the original plans. The urban landscape thus ought to be seen as an integral part of the urban transport system when discussing its long-term dynamics. This means that many new categories of actors and organisations must be taken into account. In most cities there is a rather limited number of public and private actors including leading politicians and planners, major property owners, and managers of large building companies and banks, that all have some influence over the development of the urban landscape.

In the LTS approach, the development of systems is analysed in terms of successive phases. In each phase, a certain type of actors functions as the primary system-builder and the kind of problems that they face vary from phase to phase. For example, in the early phases in the development and establishment of the electricity system, inventors and entrepreneurs such as Thomas Edison were the key system-builders, and their main challenge was to develop a system that "worked" and to assure potential customers that it was reliable. In later phases power company managers and financiers were the system-builders, and they often had to face the various technical or institutional problems that arose when the system grew. And as

you remember they were anxious to achieve a high load factor using a variety of means.
Urban transport systems are different. There is no distinctive innovation phase. There has always been some kind of transport system, and when new systems have been introduced they have functioned in parallel with older systems. And the development of the urban landscape and transport infrastructure is formed by a whole network of actors rather than by a few key system builders. Nevertheless, I will present an outline of four phases of urban transport systems in the Western world.

The first phase is the pre-industrial city, when pedestrians and horses pulling carts filled the streets. The cities were densely built, and the streets were often congested, especially on market days. The second phase is the industrial city in the second half of the 19th century. The railways had revolutionised inter-urban transport, but there were no new systems for transport within cities. Most people still had to walk, and the cities therefore remained dense and the streets became more and more congested. However, radical solutions were developed, using the space under the streets. Water and sewage pipes freed the streets from a great deal of bulky transport and later underground railways also made passenger transport possible beneath the streets.

The third phase is the early suburban city. Now solutions to the overcrowded cities were sought by organising efficient transport systems to new suburbs outside the dense city. Trams, trains and buses made this possible. Not only dwellings were built in the suburbs; more and more factories, offices and shops were located there as well, reachable by public transportation. But the different categories of activities were located in separate places, according to a new, emerging city planning ideal of functional separation. The fourth phase is suburbanisation based on the motor car, which made it possible to live further away not only from the city but also from suburban centres, and this contributed to urban sprawl over vast areas. More roads had to be built, particularly in the peripheral zones of the city regions, in order to cope with growing traffic volumes and new shopping centres and workplaces emerged along these roads, accessible only to households with cars. These developments stimulated more and more people to go by car, and in many urban regions this has led to a vicious circle of increasing traffic congestion.

Within these crude phases there is considerable variation among different cities. These differences are partly due to the character of the actor networks of each city. In cities where these actors have been able to co-operate and co-ordinate their activities, they have achieved far-reaching changes in the landscape and the transport networks according to common plans and visions. If they had not been able to co-operate the changes in the landscape would have been more piecemeal and haphazard in character. The former case, when there is an efficient and well-coordinated network of actors, is what I call a city-building regime. A clear example of a city with a strong city-building regime is Stockholm in the middle of the 20th century. It succeeded in building an underground system and a whole series of new suburbs along the underground lines.

In his analysis of the development of electricity systems Thomas P. Hughes introduced another important concept, namely momentum. This concept describes how these systems gradually acquire a kind of inherent development in a certain direction. This is due both to the growing mass of material equipment of a certain kind, but also to the values and expectations of the most influential actors in the system. Also in the development of urban transport systems including the urban landscape, there is very clear momentum. The parallel development of the networks of roads and railways and of dwellings, workplaces, schools, shops etc. mutually reinforce each other. It is very unusual that existing roads or railways are removed. Buildings may be torn down but are often replaced with other buildings on the same spot.

One can also describe this inertia with the concept of path dependency, a concept often used by institutional economists. In many fields of human activity the choice of path at an early stage will strongly influence further developments. This certainly holds for urban transport systems. Here there are many cases where it is often almost impossible to regret and undo such an early choice. It has an irreversible character. A consequence of this is that it is essential to try to do things right from the beginning. To have a very long-term perspective when making decisions that will influence the urban landscape and to strive for solutions that seem to be as robust as possible for the future.

Challenges for the future

I will now turn to the third and final part of my contribution and discuss some of the issues and challenges facing urban transport systems today. In my opinion, the major challenge for the future is to learn how to use existing roads and railways much more efficiently than we do today. Until now, solutions to congestion problems have mainly been to increase road and rail capacity, by constructing new networks, first under ground, then out of the city, and then on the periphery. I believe this kind of development has come to an end.

Firstly, there is a great scarcity of land in urban areas. If more roads are built in dense city areas, there is not much left of the city. Secondly, it is extremely expensive to build and rebuild in these areas. I suppose that many of you are familiar with the so-called Big Dig project in Boston, where a highway on pillars right over the downtown area is being replaced by a highway in a tunnel. It is only a matter of a few miles but the cost of the project is astronomical, somewhere in the order of 20 billion dollars. To learn how to use existing roads and railways more efficiently is thus a challenge, or phrased differently, to *increase the load factor*. In fact, I see huge potential, once this way of perceiving the problem has been widely accepted by the city building regimes and the public. The main difficulty at present is to gain broad acceptance for this problem formulation.

A first step towards increasing the load factor is to introduce congestion charges. These charges should also take account of the number of people in a car. There is huge potential to increase the load factor by filling the cars with passengers. We all know that it has been difficult to organise this in the past. But the widespread use of mobile phones opens up radically new potential for co-ordination. We simply need to watch the way teenagers organise their encounters using mobile phones, without making any appointments before hand. With sufficiently strong incentives, these possibilities could well be exploited within the coming decades.

Another challenge is to develop much closer co-ordination in the urban transport system as a whole, primarily in order to overcome the gulf between public transport and private cars. The aim should be to make it much easier to change mode and to use bicycles, cars and public transport in more efficient combinations. A first step towards this could be to develop a more comprehensive way of pricing urban transport. For example, an argument against congestion charges on cars in cities has been that public transport would not have the capacity to transport former car drivers. But if congestion pricing is first introduced in public transport, those public transport passengers who have a certain degree of flexibility would choose low charge times for their trips, thereby making more capacity available at peak hours.

There is one last lesson that can be drawn from electricity systems. In recent years utility managers have discovered that it is often cheaper to help customers use their electricity more efficiently than to build new generating capacity. This "demand side management" also offers enormous potential in urban transport. It is essential to realise that it is not mobility as such that the inhabitants in a city strive for. It is accessibility – to jobs, schools, services, etc. Improved accessibility can sometimes be achieved by a higher level of mobility, but it can also be reached by rearranging and relocating different kinds of functions and activities in relation to housing areas.

Chapter 3. The Meaning of the Words 'Complex' and 'Complexity'

Emin Tengström

Introduction

Two key words in this anthology are 'complex' and 'complexity'. These words can be used both as everyday words and as scientific terms. The aim of my contribution is to present some reflections on the two words in order to draw attention to their varying meanings. I am – on the other hand – not going to recommend a specific meaning.

What do we mean by the words 'complex' and 'complexity' in ordinary English? In the *Cambridge International Dictionary of English* (1995) 'complex' is defined as 'having many parts' (e.g. a 'complex structure' or a 'complex procedure'). The word can also be defined as 'difficult to understand or find answers to' (e.g. 'a complex issue or problem to which there is no straightforward answer'). In the case of the word 'complexity' two examples are given: 1/ 'there is a problem of great complexity' and 2/ 'there are a lot of complexities surrounding this issue'.

In the *Concise Oxford English Dictionary* (the large edition of 1999), the word 'complex' is defined as 1/ 'consisting of many different and connected parts' and 2/ 'not easy to analyse or understand: complicated or intricate'.

The word 'complex' can, as I said, also be used as a scientific term. To use a word as a scientific term means that the manifold and vague connotations of the word have been reduced or eliminated and the denotation of the word has been specified. As everybody knows, the term 'complex' has specific and very different meanings in disciplines such as chemistry and psychology. 'Complex' and 'complexity' can, however, also be used as scientific terms in theories that deal with systems.

To look on a transport system as a complex system

A complex system is usually composed of several interacting elements. The changes in such a system are extremely difficult to predict. The reason is that a system of this kind is often characterised by non-linear developments. The consequence of this fact is that even unimportant factors may produce great changes in the system. Changes in a certain system may even be chaotic.

Peter Allen who is Professor of Evolutionary Complex Systems, defines 'complexity' in the following way:

> There are "two basic reasons for the complexity in a given situation. Either is the complexity the result of many interconnected parts where the connections are known, or it is the outcome of non-linear interactions with bifurcation points that may result in a multitude of outcomes and creative and surprising responses. Complexity of the first kind only needs more computer power to unravel while the second type needs new views and approaches that the positivistic paradigm cannot contribute with."
> (Allen 2000:79)

This use of the scientific term 'complexity' is frequent in mathematics, physics and chemistry but is sometimes also applied to man-made systems, such as transport systems. Here, I quote professor Gunilla Jönson, who regards systems for the transportation of goods as 'Complex Adaptive Systems'. Such a system "consists of several agents that act in correlation and interdependence to each other. This is a dynamic process where the influence between the agents changes over time." She also regards such systems as "open dynamic systems that continually exchange information and energy with the surrounding environment", and she argues that "the companies exchange information with companies outside their primary field and they gain energy and information from consultants, new employees, and universities. To capture the behaviour of complex systems one has to take a holistic approach in order to realize that they are open dynamic systems "(Jönson 2003).

The complexity of urban transport development

A preliminary discussion of the complexity of urban transport development will probably be carried out mostly in ordinary English. That means that such a development is 'difficult to understand' and represents an issue 'to which there is no straightforward answer'. It is also justified to state that urban transport development is 'a problem of great complexity' or that 'there are a lot of complexities surrounding this issue'.

What do we mean by saying this? I think we claim that the transport situation in many cities is characterised by a variety of interlinked problems. The current problems are not only a question of reduced efficiency and increasing congestion. It is also a question of a transport system that is unsafe for its users. The noise of the vehicles and their emissions of various chemical substances have a negative impact on the local environment and on the health of the urban population. At the same time, the emissions of carbon dioxide contribute substantially to an increased greenhouse effect on the global level. Besides this, the future energy supply of the motorised global transport system seems to be far from guaranteed. In the perspective of social justice, it can also be said that the motorised mobility of urban citizens is today often distributed in an unequal way, which represents a serious social problem. Finally, there is also a clash between transportation of people and goods on the one hand and the aesthetic values and liveability of the urban environment on the other.

In this anthology, we will learn about the complexity of urban transport development through examples from various parts of the world.

How to cope with the complexity of urban transport development?

But there is no use in lamenting over the state of things. The discussion should focus on how to cope with the complexity of urban transport development. I think we could start from the assumption that any successful strategy to cope with the problems of urban transport must be regarded as a very complex procedure. In ordinary English this means a procedure that 'consists of many different but connected parts', a procedure that is 'complicated' or even 'intricate'. It can even be stated that a successful change process probably needs to be a messy procedure. It cannot be seen as a simple rational problem-solving process.

The reason for this statement is that the procedure for coping with the complexity is associated with many different actors, structures and processes. What actors are important to an attempt to reduce current transport problems significantly? Are these actors able to co-operate? If so, will they be hampered by the structures of which they themselves are an essential part? If this is the case, are the actors able to influence these structures? And if they are, are they able to initiate a successful implementation process of their ideas and intentions? How will they overcome the various barriers that will probably emerge during the process? How will they respond to and handle the inevitable, unforeseen (negative) surprises that will come because we are dealing with non-linear and thus not fully predictable systems?

One important question is about the possibility of an actor exerting strong leadership to handle the complexity of the development of a transport system. Will it be possible under his/her guidance to take strategic decisions of a political nature, where various actors seek to satisfy their personal and institutional need? Or will the entire process end in failure? Does the leader have the self-confidence and courage to change the original plans if necessary to compensate for negative surprises?

Against this background it is reasonable to think about possible failures. The market seems to be unable to reduce the problems of an urban transport system significantly, as market actors are seldom able to act on the level of the urban transport system. It is therefore common today to talk about 'market failures' in connection with the complexity of urban transport development. National and local governments have not been particularly successful either. It is thus also reasonable to talk about 'government failures' as a consequence of the complexity of handling the change process. It is also justified to talk about the inability of key actors (political as well as economic) to co-operate in meeting the problems of urban transport and its development. In this context I have suggested that one could talk about 'interaction failures'.

One could possibly also talk about 'public acceptance failures' as a factor behind these other failures. The individual citizen is seldom concerned with the complex problems of the transport system and the possible solutions to these problems. They are more worried about their own mobility and what it costs in money and time. The citizens may – both in their capacity as consumers and voters – be unwilling to accept and support necessary changes in the urban transport system initiated by a coalition of political and market actors. The individuals of an urban population must therefore be seen as the key actors in the process that cannot only be top-down but also bottom-up.

Even if it is natural to start by mostly talking about the complexity of urban transport development in terms of ordinary English, one should not forget the possibility of looking at the transport system as a complex system in the scientific terms indicated above. To analyse transport systems in terms of complex systems is a task for future research. One starting point for doing this is offered by the theory of Large Technical Systems described by Arne Kaijser in the preceding chapter.

This is one of the current theories in social science that have been developed in order to explain the dynamics of technical or, more correctly, techno-social systems. Such systems are characterised by three steps: establishment, expansion and stagnation. They are also characterised by certain inertia; why radical changes in the system are avoided. However, it seems reasonable to ask if there might be a fourth stage in the case of urban transport systems: a fundamental crisis.

Summing up these reflections, I think that the discussion on how to cope with urban transport development could profit from using the words and terms 'complex' and 'complexity'. These words will, whatever the case, be difficult to avoid.

References

Allen, P. M. 2000, "Knowledge, Ignorance and Learning", *Emergence*, vol. 2, no. 4, pp. 78-103.

Jönson, G. 2003, "Practical modelling of Urban Goods Transports", *Vinnova* transport call 2003

see also Nilsson, F, 2003, "Ontological and epistemological considerations in logistics research - Introducing complexity theory into the logistics discipline, *lic thesis*, Department of Design Sciences, Packaging Logistics, Lund University

Section two: Urban Transport Systems Today: a Variety of Complexity

Introduction

The first and basic condition for coping with urban transport development is to have an idea of what the problems look like in different cities all over the world. Therefore, this part of the anthology will focus on the current problems of urban transport in a number of cities in various parts of the world.

As was indicated in section one, any urban transport arrangement can be looked upon as a complex system (where the word 'complex' may be interpreted in different ways). This complexity varies from city to city. The examples are taken from developed as well as from developing countries to illustrate this variety.

In the first contributions, public transport systems are in the forefront of the discussion. This transport mode represents only a part of the entire urban transport system (it is a "subsystem") but it is important for two reasons. Its primary role is to offer a possibility of moving around to those who – for various reasons - cannot move in a more individual way (motorized or not). The existence, structure and quality of public transport therefore influence the distribution of social utilities. Secondly, increased use of public transport - instead of individual motor vehicles - is a way of reducing congestion, energy consumption and the amount of emissions having a negative effect on local and regional environment as well as on global climate.

In the first chapter, Paul Mees compares public transport in two fairly dispersed cities, Melbourne in Australia and Toronto in Canada, both of them situated in developed countries. Toronto can be regarded as a model for Melbourne. Why? Paul Mees' answer is that in Toronto there is a centralised planning institution determining the structure of the public transport. In Melbourne there is no such organising unit but only free competition between a number of operators. The result is a less well functioning public transport system with weak integration - if any- of the different transport modes.

In the next chapter, the same theme is treated but on a general level. Hans Örn introduces a distinction between 'the service structure' (being a coordinated network or a fragmented network) and 'the organizational structure' (monopoly or competition). He finds that the complexity of the public transport system can be handled in three, quite different, ways. One is what he regards as "The classical European Model". Here a coordinated network has been run by the city administration as a monopoly (for instance in cities as Stockholm, London and Moscow).

The opposite way of handling the complexity of public transport is to leave it to private operators who – in a situation of competition - decide independently about their routes and services. This has the consequence of making the network fragmented and less coordinated. This way of handling the public transport system (earlier recommended by the World Bank) has been applied in cities like Dhaka, Nairobi and Lima. It has also been introduced in the UK in the middle of the 1980s during the Thatcher era.

A third way of organising the public transport system is a kind of controlled competition based on a balance between the private and the public sector. A public transport authority is responsible for 1) route network planning, 2) provision of infrastructure, 3) concessions/ subcontracting and 4) monitoring. This system has been introduced in Swedish cities such as Stockholm and Göteborg (Sweden's second city), and is the recommended approach in the European Union.

In the final part of his chapter, Hans Örn takes Kingston, Jamaica, as an example of how complex the reform process can be in a developing country. After independence the British private bus monopoly was taken over and replaced by a public monopoly. Following the advice of the World Bank, public transport was later privatised and deregulated. As that situation became intolerable as a consequence of fierce competition, an attempt was made to introduce a system of controlled competition similar to the Swedish model. This attempt failed as the deregulated concept was too deeply rooted. Kingston has now returned to the concept of a state owned monopoly. Plans have, however, been produced to split up the monopoly in the future and establish the controlled competition concept. In conclusion, Mr Örn suggests that international advisors to developing countries have been too dogmatic in their faith in market forces and have neglected the role of the public sector.

In the third chapter, Ahmad Kanyama *et al* analyse the transport development in the city of Dar-Es-Salaam in Tanzania against a background of how African cities have been changing during the last years. There is rapid urbanisation with the population in African cities expected to grow from about 200 million in 1990 to more than 450 million in 2010. This leads to very dispersed cities as most dwellings are privately owned one-family houses. This situation creates – in its turn - problems for the population to move around to work, to schools, to hospitals, etc. Apart from deteriorating services to the public, the situation has led to traffic congestion, air pollution and many road accidents.

The city of Dar-Es-Salaam is representative of this development. Its population has increased from 356.000 in 1967 to 2.5 million in 2002 (at an average yearly rate of 4.5%). The city spatial expansion has been growing at an average of 7.2% per year. The present transport system consisting of minibus para-transit vehicles (run by a number of private firms) is inadequate and expensive for the poor. The public transport system can, according to the authors, only be improved, 1) if a link is established between land use and transport planning, 2) if there is some coordinated planning of the activities of the transport operators and 3) if there is a more adequate institutional coordination among government sectors. The successful introduction of the Bus Rapid Transit system (BRT and extended facilities for non-motorized transport) are seen as ways of managing public transport in Dar-es-Salam.

In the next chapter, Liya Liu analyses current urban transport problems in Chinese cities. She describes China's remarkable economic growth and development since 1978 still continuing at a very high level in terms of both GDP and by other measures of growth. This economic boom has led to rapid urbanization. While China's urban population exceeded 520 million in 2003, it is expected to reach more than 710 million in 2020. Such a sustained high level of economic activity has resulted in intensified road traffic of vehicles carrying freight and passengers.

The domestic production of such vehicles is seen both as an important part of the economic boom and as a necessary means for solving transportation problems throughout China as a whole and in its urban centers in particular. This development has, in its turn, led to an increased energy demand particularly in China's transport sector, highlighted by the fact that China has been a net oil importer since 1993. China has therefore become very vulnerable to the uncertainties of the world oil market with respect to both availability and price. China is today the world's second largest energy consumer after the United States.

Chinese urban transport systems are today characterised by:

1. Inadequate public passenger transport
2. A rapid increase in vehicle mobility
3. Rapid urban sprawl
4. Serious traffic congestion in many cities
5. Inadequate planning of the development of urban transport systems

There are a number of negative consequences of the current situation in China's urban transport systems that are triggered by rapid economic growth:

1. Greater energy insecurity caused by increased energy consumption by motorised vehicles
2. Severe air and noise pollution in many cities, with vehicle emissions causing premature deaths and resulting in a number of other

adverse health effects. All these factors also contribute to the risks of climate change

3. High traffic casualties. In 2002 China suffered 773,137 traffic accidents, causing 109,381 deaths and 526,074 injuries

There is no panacea to reduce these problems. Building new road infrastructures does not offer a way out. A greater use of vehicles is no solution. What to do then? Today there is a limited capability to address the problems of urban transport in China. Many city governments have not yet acquired either sufficient experience or the skills to deal adequately with the complexity of urban transport growth.

Liya Liu suggests that one way to reduce the problems in many cities is to introduce a Bus Rapid Transit (BRT) system. This system has a number of advantages, as has been demonstrated in Curitiba and Bogotá in South America. In China several initiatives have already been taken to promote the introduction of BRT. It is, however, clear that the introduction of the system is not by itself enough to solve all the problems. It must be supplemented by policies that establish public transport priorities not only for passenger carrying vehicles but also for the movement of goods by trucks taking into account other issues such as energy consumption, pollution and road safety.

In the last chapter of the second section a different approach is chosen. While the focus of the preceding chapters is on the role of public transport, Marie Thynell discusses urban mobility and its relation to societal development in a global perspective. Her main focus is on what is called in American English the "automobilisation of transport systems". The background of her study is the well-known negative impact of motorized mobility in cities, where she emphasizes the social division between those who have access to motorized mobility and those who have not.

She then presents a recent study of how a motorized transport system was built up in two very different cities, Brasilia in Brazil and Tehran in Iran. She also compares how the current problems of urban transport are perceived and handled in two other very different cities, Rome in Italy and New Delhi in India. Repeated statements by her informants saying that the problem-solving capacity is to be found in new technology leads her to a study of the views expressed by three big automotive companies and one big oil company. She finds that these market actors mostly focus on the technical development of their products and their own production and not on the problems of the entire urban transport system.

One of the fundamental problems of urban transport development is, therefore, that different actors perceive the problems of urban transport differently. They all rely on socially constructed (but often far from rational) views of the problems and the strategies. The politics of mobility are often concentrated to the role of the car. Policies for urban mobility have, so far, favoured market interests and, at the same time, contributed to the marginalisation of the urban poor in many countries.

From these studies Marie Thynell draws the conclusion that current urban transport systems are approaching a fundamental crisis in their development, and that none of the traditional groups of actors (politicians and market actors) seem to be able to confront the crisis of urban motorized mobility. They seem unable to increase mobility, sustainability and social equity at the same time. There is, therefore, a desperate need of a new and better knowledge base for future actions. She points to the importance of studying the dynamics of change. Doing so, she stresses the role of the specific geographical, political, economic, social and cultural context of each urban transport system. Furthermore, the political and economic role of automobility is still poorly understood.

The basic question is, therefore, what kind of transport policies contribute to societal development in various urban and socio-cultural settings. At the same time it is necessary to identify and analyse relations between the local and international levels of mass motorization. And, finally, what is the inherent conflict between the demand for societal development and a modern lifestyle based on the use of the car?

Chapter 4. Responding to the Complex Transport Needs of Dispersed Wealthy Cities: a Comparison of Melbourne, Australia and Toronto, Canada

Paul Mees

Introduction

The transport problems of Melbourne are those of an affluent community characterised by high car ownership and dispersed land-use patterns. In this respect, Melbourne is similar to Toronto, but in some key respects the experiences of the two cities have diverged. An examination of this divergence can, I believe, shed some light on the policy measures needed to respond to the complexities of transport in such an urban environment.

Transport in an affluent city

Local politicians in Melbourne still boast about the fact that the city was designated 'the world's most liveable' in a survey some ten years ago. Urban incomes are high and inequality, while increasing, is less pronounced than in many British or US cities. Most households can afford cars and the city has inherited broad streets from its 19th-century planners. Public transport infrastructure is also extensive: the city boasts one of the largest urban rail networks, relative to population, in the world, and has also had the good sense or good fortune to retain its tram system.

Unfortunately, all is not quite as pleasant as surface impressions might suggest. One household in eight is without the use of a car, and many of those that do have cars find operating private vehicles a financial burden. This burden is exacerbated by the fact that rising inner-city property values are increasingly driving the less well-off to outer suburbs, which have poor or non-existent public transport. The suburban municipality of Greater Dandenong, for example, houses 132,000 people and has the lowest per capita incomes in the metropolis. There are no trams and not a single bus service that operates at all on Sundays or weekday evenings.

Excessive automobile use produces local air pollution which results in frequent 'smog alert' days, and contributes to Australia's dubious record as the world's highest per capita emitter of greenhouse gases. Meanwhile, the local media is obsessed with stories about rising obesity levels, particularly among children, which are increasingly being blamed on sedentary lifestyles characterised by little or no walking.

Because the extensive rail and tram systems are so poorly patronised, they, along with the bus services, require substantial government subsidies to keep them afloat. The Victorian State government, which is responsible for transport policy in Melbourne, attempted to solve this problem by privatising trains and trams in 1999 (the buses have always been privately run). Although optimistic predictions were made of improved services and falling subsidies, the actual result has been the withdrawal of the major private operator and a dramatic blow-out in subsidy requirements (Mees, 2002).

Policies or political correctness?

Melburnians have become cynical about public transport, which seems to lurch from one crisis to another. A not uncommon view is that the task of providing effective alternatives to the car in a community like Melbourne is simply too difficult. Many professional transport planners and media commentators support this view, arguing that the complex travel patterns of the 'post-modern' city cannot be served effectively by public transport.

The majority of the public appear to have somehow retained their optimism. When the Victorian government consulted the community about priorities for its new metropolitan strategy for Melbourne in 2001, the results surprised almost everyone (including this writer!). The most common issue raised by participants in the consultation process was the need to fix public transport, and to do so ahead of road improvements, if necessary (Mees 2003, p. 297).

The Victorian government responded in its strategy report, titled *Melbourne 2030* (DOI, 2002), with numerous strong-sounding statements about the need to give priority to sustainable transport modes and move away from automobile dependence. The report itself is liberally illustrated with colour photographs of people walking, cycling and riding trams. The strategy also contains an extensive program of freeway construction – although, interestingly, this is not set out in the transport chapter, but is instead hidden away in a chapter entitled 'A more prosperous city'.

It appears, then, that the rhetoric about sustainable transport is a form of 'political correctness' that obscures the real agenda. This is a long-established practice in Melbourne, dating back at least as far as the 1969 Melbourne Transportation Plan, which visiting British transport economist J. M. Thomson described as:

> an unconvincing piece of work. It is based on the earlier American transportation study techniques, by now thoroughly discredited, and it was presented with all the glib political clichés one has learned to distrust. The public relations document bears the labels 'Train', 'Bus', 'Tram', 'Car', in that order, and begins with a description of the railway plan, followed by those for trams and buses. Lastly mentioned are the highway proposals. At the very end are given the costs [86% for roads and car parking; 14% for public transport]... Clearly the plan is a highway plan, not – as it is called – a comprehensive transport plan (Thomson 1977, p. 137).

In the case of Melbourne 2030, the 'political correctness' conclusion is buttressed by the absence of substantive proposals for promoting a shift from the automobile to 'green' modes. The strategy proposes new bicycle paths, but the major links are to run adjacent to new freeways, which provide very poor route locations – unless the real purpose of the bike path is to enable transport planners to claim the bike path somehow negates the pro-automobile bias of the freeway. The major initiative promised for public transport is a behaviour change program called Travelsmart. While there is some evidence that marketing programs of this kind can be effective in situations where alternatives to car travel exist, or have been improved, the UK Department of Transport warned:

> Ideally the techniques should be used in the context of a wider comprehensive travel demand strategy. It would seem that they need to be thought of as an integral part of a strategy rather than as some form of 'public relations' exercise, when nothing is being done to address strategic transport priorities (DOT 2002, section 8).

Explaining the difference between words and actions

By engaging in 'political correctness' exercises, Melbourne's transport planners are acknowledging that there is widespread public support for a reorientation of transport priorities away from the car. But by failing to provide any substantive change of direction, they also reveal that they do not share the public's view.

An indication of one of the reasons for transport planners' scepticism about the potential for alternatives to the car can be found in the Melbourne 2030 report's discussion of urban density:

> [D]ue to the form of development after the Second World War, the average density of the metropolitan area at around 14.9 persons per hectare (pph) is low by international standards. Montreal has 33.8 pph, for example, and Toronto has 41.5 pph (DOI 2002, p. 60).

The choice of cities to cite is significant, as is the source of the density figures. Although no source is given in the Melbourne 2030 report, the density figures correspond precisely to those given for the three cities in 1991 by Newman and Kenworthy (1999). These authors are strongly associated in Australia with the notion that Toronto offers a positive example for Australian urban transport policy-makers. This idea is not new. Back in 1968, the authors of the Sydney Region Outline Plan stated:

> The general conclusion that can be drawn from a preliminary examination of the public transport system is that there will be an increasing need for the co-ordination of and interchange between the different modes of public transport... Much remains to be done in this area before Sydney can experience the benefits of a public transport system as good as Toronto in which bus and rail services are closely integrated, passenger transfer from one system to the other is made convenient by the existence of carefully designed interchange stations, and tickets for both systems are fully interchangeable... (SPA 1968, p. 43).

Toronto remains a model for those seeking to revive public transport in Australian cities today. Public transport patronage is certainly higher in Toronto than in Melbourne, although it is significant that the picture was reversed four decades ago (Table 4.1). Toronto's higher patronage is carried on a very much smaller heavy rail infrastructure base.

	1950	1960	1970	1980	1990	2000
Toronto	292	183	185	213	223	188
Melbourne	449	222	142	95	97	96

Table 4.1. Public transport patronage (unlinked trips) per capita, 1950-2000

Sources: Mees (2000, p. 178); Annual reports and census data for 2000/1.

Toronto is lauded as a model of 'transport/land-use integration', of the kind of high-density development, especially around railway stations, that proponents of urban consolidation advocate (e.g. Newman & Kenworthy, 1999). This seems to be the implicit argument behind the citation of the density figures in Melbourne 2030 (see above). But the 1968 Sydney plan presented Toronto as a model of something different, a fully integrated public transport system. Which of these two images of Toronto is correct? And perhaps more importantly, which of them really explains the reputed success of its public transport system?

The Newman & Kenworthy density figures for Toronto are not calculated on a comparable basis to the Melbourne figure, as can be seen easily by comparing the maps of urbanised areas for the two cities shown on pages and 229 and 375 of Kenworthy et al (1999). When calculated on a consistent basis, the densities of the two cities are much more similar than has generally been believed (Mees 2000, pp. 190-3). And the distribution of travel and economic activity is actually less favourable to public transport in Toronto, because that city's growth was influenced mainly by its freeway system, which predates, and is more extensive than, the rail system. The high-rise apartments found at a number of subway stations in Toronto have been given a great deal of prominence in urban consolidation debates, but in fact there are only half a dozen stations like this in the whole city.

So if I am correct here and Toronto cannot teach Melbourne much about transit/land-use integration (not because it is unimportant, but because Toronto's performance has been exaggerated), what can it teach us?

Transport policy differences between Melbourne and Toronto

The history of Melbourne and Toronto has many common features, such as the British colonial background and the transforming effects of post-world War II immigration. Even transport planning history shows parallels.

Both Melbourne and Toronto faced public transport crises at the end of the First World War. Each city had franchised a private firm to run its tram system, and the franchises had expired following periods of public dissatisfaction with private operation. In 1919, Melbourne set up a public agency, the Melbourne and Metropolitan Tramways

Board, to run its trams. The MMTB competed with the existing government-run suburban railway system, and later with private bus operators who commenced in the 1920s and operated according to a deregulated 'free-for-all' (Mees 2000, chapter 9). Because even the government agencies competed against one another, Melbourne was something of a model of the 'free market' approach to urban public transport.

Toronto, by contrast, has been a model of centralised planning by a single public authority. The Toronto Transportation (later Transit) Commission was established in 1921 to service the City of Toronto and expanded into the growing suburbs in 1954. The TTC was given a monopoly of all forms of urban public transport except taxis: no competition was permitted. There were no suburban railways in 1921: when these were opened from the 1950s onwards, they were developed by the TTC.

Both the MMTB and the TTC appear to have been efficiently-managed organisations in the 1920s and 1930s, but they operated differently in two critical respects. Firstly, the TTC was a multi-modal operator from the very beginning, using buses to supplement trams in outer areas and on cross-city routes, and providing multi-modal fares (with free transfers) and integrated timetables. When the first subways were built in the 1950s, the integration theme continued: routes and stations were designed to maximise the opportunities for intermodal transfer. Secondly, as a municipal body the TTC was charged with serving all the residents of the City of Toronto, and established early on the principle of minimum service standards across the whole service territory. After 1954, the whole Municipality of Metropolitan Toronto became the TTC's service area. Initially, the TTC attempted to provide a lower service standard to the suburbs, on the basis that population densities were too low, but this policy was reversed by suburban representatives on the TTC's managing board. In 1963, suburban service levels were upgraded to the same standard as the City: interestingly, it was at this point that per capita patronage finally turned the corner and began to increase (Mees 2000, chapter 9).

By contrast, in Melbourne, the different public transport operators treated one another as competitors: intermodal integration was virtually non-existent, and it was common for services to parallel one another. The level of service available to different areas of the city was a result of historical accident, rather than planning, and varied wildly and irrationally. Those living close to tram routes were better served than those living along bus routes, even in areas with similar densities and urban forms.

The divergent public transport service models determined the way in which the two cities responded to the rapid increase in car ownership that began in the 1950s. In Toronto, there was a planned 'counterattack' by public transport, as bus services were expanded into the suburbs. In the 1970s, following the cancellation of some contentious freeway projects, the subway, which originally was to be confined to the inner city, followed into the suburbs. In Melbourne, by contrast, public transport simply entered an unplanned spiral of falling patronage, service reductions and fare increases. Like Toronto, Melbourne was convulsed by debate over freeways in the 1970s, and the State government promised a reorientation of priorities toward public transport. But the problem in Melbourne was that nobody knew how to plan or integrate public transport, and the pattern of intermodal competition and patronage decline continued.

The 1990s: the Toronto model comes unstuck

Both Victoria and Ontario elected neo-liberal State governments in the 1990s, and their policies affected public transport significantly. In Melbourne, the rail and tram systems were privatised, along with the small section of the bus system that was in government hands. In Toronto, public transport remained in municipal hands, but the Province stopped contributing to capital and operating costs. The result was a substantial reduction in service levels which, combined with fare rises, produced a sharp drop in patronage (Table 4.1) – although it remained at much higher levels than in Melbourne. Capital expansion programs were put on hold and the TTC was required to defer maintenance and equipment upgrades. The Premier rejected pleas for funding, telling the Provincial legislature that the TTC should be:

> reviewing its business practices and doing what they have done in other jurisdictions, for example, in Australia, where lots of places have contracted out... public transit. In one case, I forget whether it's Adelaide or Melbourne, they actually have 52 different entities contributing to the overall mass transit system, and it functions a heck of a lot better than the TTC (*Toronto Star*, 23/10/02).

Ironically, at the same time, politicians in Melbourne were grappling with the failure of the very privatisation policy that the Ontario Premier was praising (see Mees, 2002). The reversal of fortune for Toronto's public transport in the 1990s shows that it had been service policy, rather than urban form (which did not change significantly over the decade), which had produced the city's superior performance in earlier decades. The Premier's failure to realise this mirrors a similar failure of understanding in Melbourne.

The fortunes of public transport in Toronto may have begun to revive following a change of Provincial government in late 2003. Public dissatisfaction with cuts to transit funding was widely cited as a major factor behind the change of government.

What the two cities can teach us about solutions

Many commentators have questioned whether it is practicable to provide viable public transport in dispersed cities. Observers inspired by neo-liberal economic theory argue that the best way of providing flexible services to a population with diverse travel patterns is through the application of competition, privatisation and deregulation. Support for this idea is particularly strong in the United States, where the poor performance of publicly owned and financed systems has persuaded even many pro-transit commentators that the market is the way forward.

But the neo-liberal theorists are wrong. While the competition and planning approaches to urban public transport sound attractive at first glance, only planning works in reality. With sensible planning, it is actually possible to have 'European-style' public transport, even in dispersed urban environments.

This is the most important lesson that can be drawn from the success of the 'Toronto model' between 1950 and 1990. The reason public transport has been so much more successful in Metropolitan Toronto than in Melbourne is the different policies adopted toward its planning and operation in the two cities. Bureaucratic planning, tempered by politics, provided Metro Toronto with 'European-style' public transport that has until very recently been able to give the kind of diverse, flexible service the post-modern city needs. And the change away from this model in Toronto in the 1990s, driven largely by politicians inspired by neo-liberal ideology, has produced a declining performance that only emphasises the validity of the approach traditionally employed there.

Melbourne's uncoordinated, market-driven public transport systems have collectively proven less able to respond to the changing travel needs of a dispersed city than Metro Toronto's single, regionally planned system. Public transport operators in Melbourne have competed with one another; Metro Toronto's single operator has competed with the car.

The lessons from Melbourne and Toronto reinforce those learned in other places, like Britain and Chile (Mees 2000, chapter 4). Only central planning enables the provision of flexible travel options through a fully integrated network. This requires the following conditions:

- an integrated route structure which maximises opportunities for interchange and reduces duplication and overlap
- fast, frequent, reliable service on the trunk (rail, busway or whatever) routes

- high service levels on all routes (cross-suburban as well as radial) throughout the day and evening
- convenient, attractive and safe interchange facilities
- matching hours of operation on the different routes serving interchanges and either coordinated timetables or very frequent services
- multi-modal fares (free transfers)
- easy-to-obtain, well-presented route and timetable information covering the whole multi-modal network

The key here is planning, rather than merely regulation, an unhappy compromise which often reproduces the worst features of both the market and planned models, by protecting inefficient private operators while preventing the type of comprehensive service provision post-modern cities really require. Regulators are less accountable to the public than planners, and are much more prone to 'capture' by vested interests, particularly private transit operators – as recent experience in Melbourne shows (Mees 2002).

Public ownership is not enough by itself either: many transit systems that have never been privately run retain outdated, inflexible radial network structures and infrequent, irregular, uncoordinated timetables. Clearly more than just regional planning is required for public transport to succeed, as the negative experience of places as diverse as Canberra and Calcutta shows. The Canadian and Swiss success stories show that a particular dynamism and creative interplay with politics is needed. More research is needed to identify the factors that contribute, positively or negatively, in this area, although it is noteworthy that public transport systems that were taken into public hands as going concerns (as in Europe and most of Canada) have tended to outperform those that were only nationalised when they failed (as in the United States).

Conversely, the experience of Zurich (Mees 2000, chapter 5) suggests that central planning need not exclude private operation of individual services, as long as this occurs within an overall framework of publicly accountable planning. Competitive tendering can sometimes be a good way of keeping costs down. Coordination and planning are the keys that make good outcomes possible, however. Although they cannot guarantee success, poor outcomes are a guaranteed result of the 'market' model, especially in dispersed cities.

One clear lesson is that we cannot simply get the rules right and sit back, waiting for an invisible hand to produce good outcomes. Good urban outcomes must be planned for. This will require first-rate public transport, but as part of an overall set of policies which also includes urban planning, traffic restraint and measures to improve conditions for pedestrians and cyclists. Policies of this kind are much easier to 'sell' politically if the public can see that public transport provides a viable alternative to the car. While it has been able, until recently, to offer effective public transport, Toronto is not a model of sustainable transport, because it has not pursued policies in these latter areas to any significant extent.

With public transport itself, the critical issue is flexibility. And the key to flexibility for passengers is simplicity and predictability, not a bewildering array of constantly changing options. The latter produces confusion, not convenience. Paradoxically, to be flexible, public transport must also be rigidly predictable: perhaps the best analogy is with the road system, rather than with cars themselves. Flexible public transport can take advantage of the increasing diversity of travel patterns in post-modern cities to produce a more even flow of passengers throughout the day, reducing the economic costs of heavy 'peaking'. It can also provide a genuine alternative to multiple car ownership, reducing the financial burden on struggling households, reducing car ownership levels generally and providing an environment in which it becomes possible for walking to play a major role in local travel.

References

DOI (Department of Infrastructure, Victoria) (2002): *Melbourne 2030: Planning for Sustainable Growth*, DOI, Melbourne.

DOT (Department of Transport, UK) (2002): *A Review of the Effectiveness of Personalised Journey Planning Techniques*, London.

Kenworthy, J., Laube, F. and others (1999): *An International Sourcebook of Automobile Dependence in Cities 1960-1990*,University of Colorado Press, Boulder.

Mees, P. (2003): 'Paterson's Curse: the Attempt to Revive Metropolitan Planning in Melbourne', *Urban Policy & Research* 21(3), pp. 289-301.

Mees, P. (2002): 'Privatisation of Public Transport in Melbourne', *Proceedings of 25th Australasian Transport Research Forum*, Bureau of Transport Economics, Canberra (CD-ROM).

Mees, P. (2000): *A Very Public Solution: Transport in the Dispersed City*, Melbourne University Press, Melbourne.

Newman, P. & Kenworthy, J. (1999): *Sustainability and Cities: Overcoming Automobile Dependence*, Island Press, Washington DC.

Thomson, J. M. (1977): *Great Cities and Their Traffic*, Gollancz, London.

SPA (State Planning Authority of NSW) (1968): *Sydney Region: Outline Plan 1970-2000 AD*, Sydney.

Chapter 5. Urban Public Transport in an International Perspective

Hans Örn

Introduction

One of the striking characteristics of our time has been the rapid urbanization process in many of the so-called developing countries – a process that is still ongoing. When the British left India, Dhaka was a town of some 400,000 inhabitants – about the same as Göteborg today. New York, London and Paris were the big cities then. 50 years later, Dhaka belongs to the world's new megacities with over 10 million inhabitants. Such rapid growth naturally puts pressure on

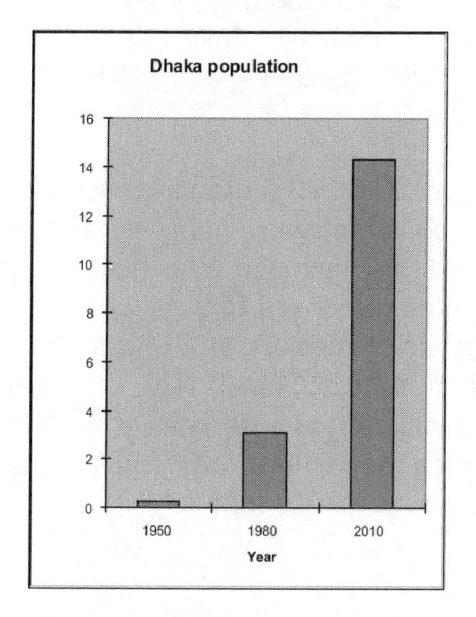

Figure 5.1. Dhaka population

all city functions, not least transportation, which is the circulation system of the urban organism.

Everyone agrees that public transport is an essential component, particularly in low-income cities. Advisors and consultants have an important role to play in assisting in the development of sustainable public transport systems. However, the discussion often focuses on technology, such as whether rail or bus is the "best" solution. In reality, the decisive issues are often of a different nature such as:

- Roles of the private and the public sector?
- Social service or business?
- Should politicians or market forces decide?
- Regulation or deregulation?
- Is it small that is beautiful – or big?

Organizational models for public transport

The South-West Sector – the "Classical European Model"

A public transport system can be described in two dimensions, one axis being the network and service structure, the other being the organizational structure.

The service structure axis ranges from a "coordinated network" to a "fragmented network". The "coordinated network" is an integrated combination of routes and services, operated by a fleet of vehicles under organized conditions in an integrated fare system. The "fragmented network" is in reality not a network at all but a number of individual routes, separated from each other. The organizational structure axis ranges from a monopoly to a situation with "full competition". This creates four different combinations which can be seen as representing different possible scenarios for a city.

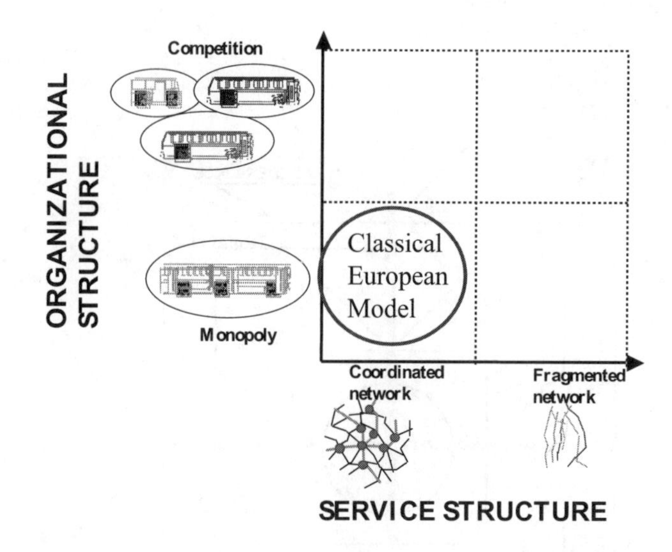

Figure 5.2. Public transport structure – the "Classical European Model"

The "classical European model" is represented by the South-West area. This is the predominant system that European cities - East or West - have applied for many years, and by and large it has served them well! The public transport system is one single entity (even though it can be composed of different technical sub-systems such as bus or rail), and can, therefore, provide an integrated and coordinated service to the whole city. The flipside of the coin is a public sector monopoly that is often neither cost-effective nor market-oriented.

A coordinated route network

One characteristic feature of an efficient public transport system is a differentiated route network. In an integrated system, different route types and different vehicle types have different roles to play.

In cities with heavy rail systems, buses normally act as feeders. In other cities, high-capacity buses form trunk lines, and smaller buses are feeders. It is particularly important that trunk lines are given high mobility, and designated bus lanes are, therefore, often designed along such corridors.

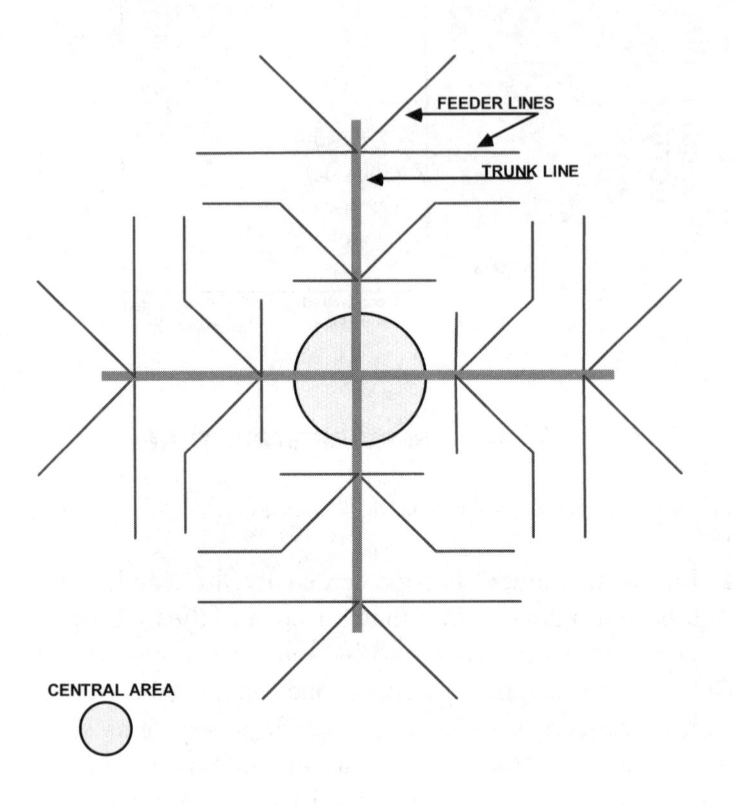

Figure 5.3. The trunk line/feeder line network

Designated bus lanes

An efficient public transport system requires mobility, so that a reasonably high operating speed can be achieved. This is for two reasons: (1) to ensure the best possible economic efficiency and (2) to make the system attractive. This makes it necessary to allocate separate space to buses and trams.

In old European cities, however, road space is often limited due to the existence of a historic city centre. It is, therefore, seldom possible to "build away" the problems with new infrastructure. Instead, existing roads and streets have to be used as efficiently as possible, and this makes it necessary to design bus lanes carefully with as little effect as possible on other traffic.

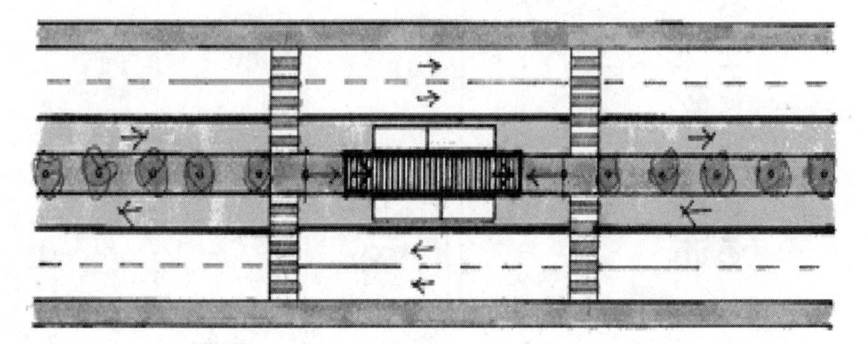

Figure 5.4. Mid road bus lane

For high-capacity bus and tramway lines, mid-road lanes are often preferred, but many other solutions have to be considered, such as kerb-side lanes. It is important to realize that there are no tailor-made solutions - each city must find its own optimal system.

Efficient and attractive transfer terminals

In an integrated system, transfers are necessary. This is particularly true when a part of the system consists of high-capacity corridors which, of necessity, cannot cover the whole city. Transfer stations should be designed to offer passengers comfort and also to speed up the transfer process, so that economy of operation is preserved.

The deregulated model

Cities like Stockholm, London and Moscow have shared the same general concept. They have considered public transport to be a service function comparable to electricity, water supply or waste collection - all of which are run as branches of the city administration.

In the opposite corner of the graph, the North-East segment, the concept of public transport is radically different. Here public transport is not considered as a social service or a city function, but as a business like any other.

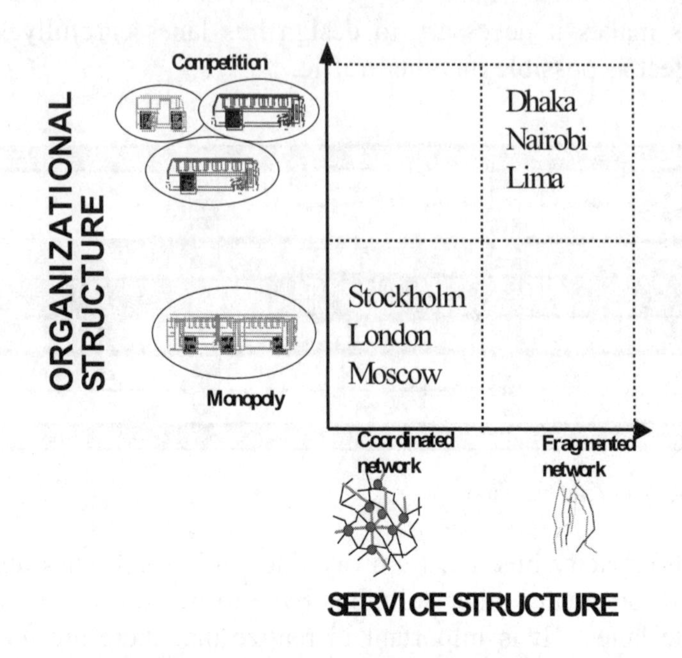

Figure 5.5. Public transport structure – deregulation

Here there is no monopoly, but free, and sometimes fierce, competition between different providers of public transport services. It is every man for himself, and as a consequence there is no interest in coordinating services. Each minibus is an economic unit and competes with all others. Each route is independent of other routes, and transfers will cost an extra fare payment.

The system typically offers the passenger a frequent and much-appreciated service along one route from the outskirts to the city centre, but does not offer area coverage or accessibility to the whole city. The concept of free competition and a fragmented network includes many cities in developing countries, such as Dhaka, Nairobi and Lima.

Leaving the "Classical European Model"

Privatization in the UK

The European model faced an increasing dilemma as car ownership increased and the use of public transport decreased. This eroded the profitability of the sector, and the part of the operations that had to be financed by taxes instead of fare revenues increased alarmingly during the latter part of the last century.

The UK became the first European nation to abandon the "Classical European Model" as Mrs Thatcher launched a privatization and deregulation process for the public transport sector in the mid 1980s.

Today 80% of bus services in the UK are run by competing operators, allowed to plan their own routes and services and to set their own fares. The requirement is that they register their planned services with the Traffic Commissioners, specifying their intended routes, timetables and fares.

20% of bus transport in the UK is still financed by the public sector for social reasons, and these services are subcontracted from private operators by the local Public Transport Executive.

Deregulation in developing countries

The British reform occurred at a time when the "cold war" was at a peak and when public transport was seen as one ideological battlefield. Also, Mrs Thatcher's advisor, Sir Alan Walters, had had a successful background at the World Bank, one of his specialities being to question the concept of large bus companies and instead promote the individually operated minibus concept.

In many developing countries monopoly bus companies existed, sometimes a heritage from colonial times, sometimes modelled on the European system or the Soviet Union, then still a conceivable alternative to the market economy concept. Many of these bus companies were badly run, mismanaged and heavily subsidized to no visible effect.

Deregulation in developing countries

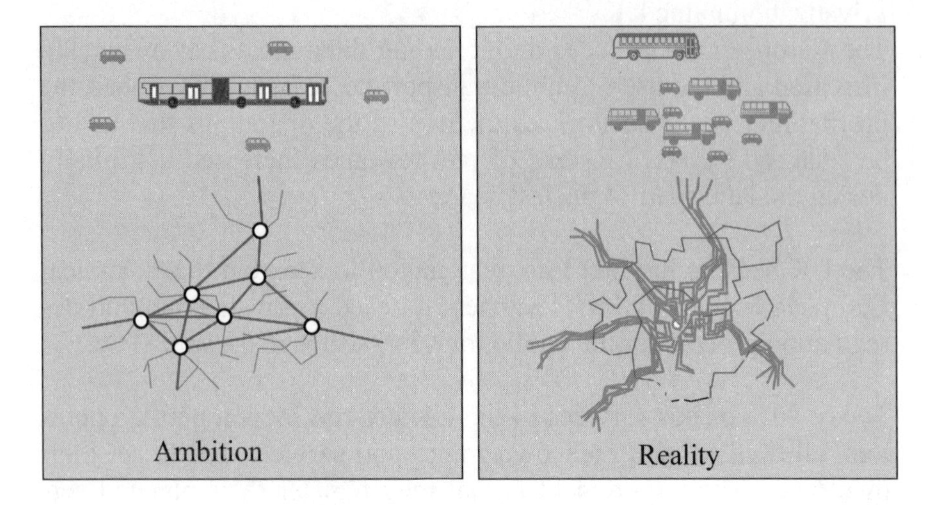

Figure 5.6. Deregulation: ambition and reality

The natural concept for many Western advisors and consultants at the time was to advocate deregulation, and to invite the private sector to operate minibuses. This was seen as a useful complement to regular bus services, and it was often believed that a trunk line/feeder line system would naturally emerge.

In reality the minibus concept proved capable of out-competing bus services, in particular in cities with little monitoring and control. Instead of being a complement and a feeder system, minibuses soon took over the profitable large passenger flows in the city centres leaving the government buses to become an unappreciated social service in the outskirts.

Fragmentation of services – effect on road space efficiency

The free competition concept in combination with high unemployment appears naturally to lead to situations with very large numbers of small, low-cost vehicles. In its extreme form, the free competition/fragmented network concept can lead to a quite inefficient use of the road and street network.

When public transport takes the form of many small vehicles where street capacity is limited, then it is obviously no longer justified on the grounds that it relieves congestion. In fact, while public transport is promoted in the West because it is hoped that people will leave their cars at home, in many developing countries the best potential for improving congestion as well as the environment is if people can be shifted from small public transport vehicles to large buses.

The Controlled Competition Model

Moving North – the Swedish reform

In 1989 Sweden took a decisive step in order to reform and restructure its public transport sector. The idea of a public sector monopoly for public transport had become obsolete - not least because of the irreversible change in the ideological climate that came about as a result of the fall of the Soviet Union. "Competition" now became a slogan, as it was considered that the efficiency and drive of the private sector would help reduce the large subsidies.

But Sweden did not want to give up the positive side of the old system - namely the coordinated network and service structure that was by now required by the voters. Thus public transport had to be reformed, but not destroyed!

The resulting policy direction was clear; Swedish cities like Stockholm should move upwards from the South-West to the North-West sector in the "policy graph". They should not be allowed to move North-East and introduce unlimited competition with a resulting fragmented network and service structure.

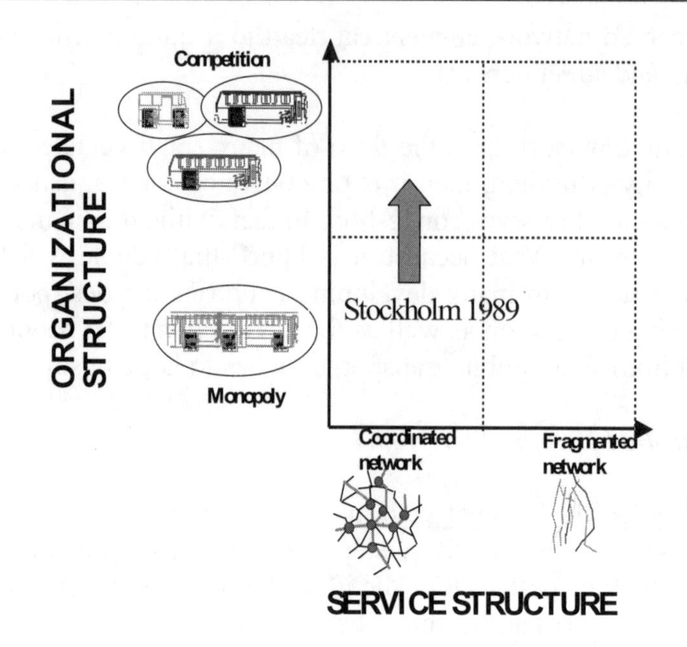

Figure 5.7. Public transport reform – "going North"

But how could this be done?

New roles for the public and the private sector

The solution to this problem was a new definition of the responsibilities for the public and the private sector, and a new role for the old public transport companies.

The public sector now created a separate unit, a Transport Authority, with the responsibility for (i) planning the route network in the best interests of the city and the passengers, (ii) providing the necessary infrastructure such as bus stops, separate busways and terminals, (iii) negotiating with and subcontracting operators for routes or route packages, and (iv) monitoring and controlling the performance of such operators. Planning thus remains a public sector responsibility in Swedish cities.

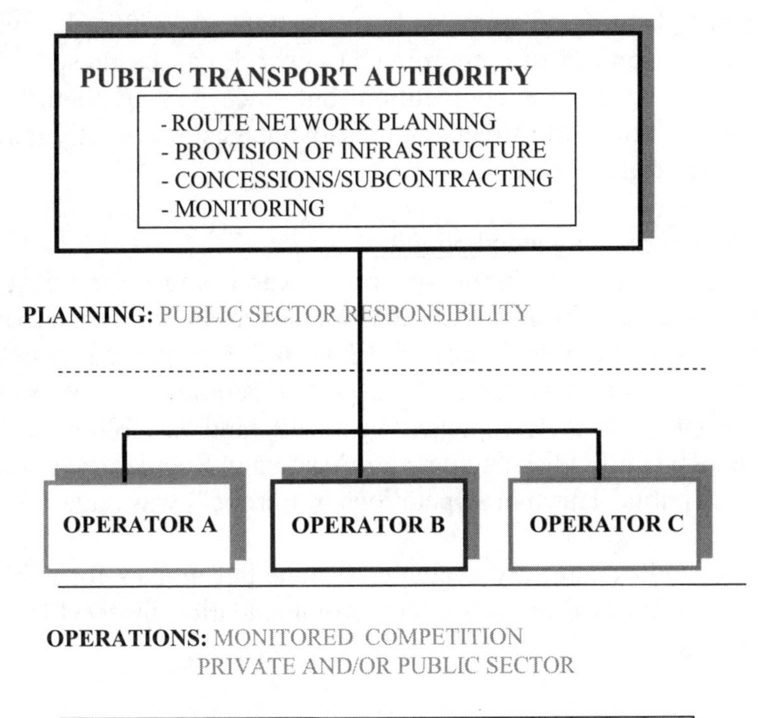

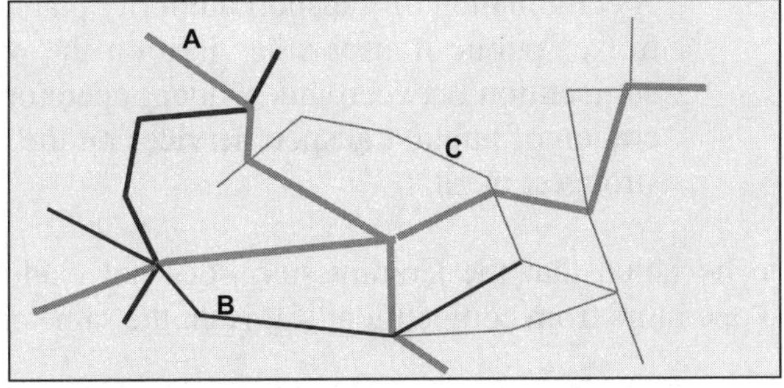

Figure 5.8. Balance between private and public sector

The operation of bus and rail services will, however, no longer necessarily be undertaken by publicly owned companies. Route packages are designed for, and subject to, competitive bidding. The operator able to offer the best price will be awarded a contract for a specific period of time.

Since planning of routes and services remains within the public sector, the concept of a coordinated network can be preserved. At the same time there is competition, but since it is competition *for* the market, not in the streets, the disadvantages of a totally deregulated system can be avoided.

The EU recommended policy…

In the European Union, specific research and policy development projects have been undertaken to analyze the issue of urban public transport. An initial project, ISOTOPE ("Improved Structure and Organisation for urban Transport Operations of Passengers in Europe"), was completed in 1997 and a follow-up project, MARETOPE ("Managing and Assessing Regulatory Evolution in local public Transport Operations in Europe") was undertaken later.

The studies carefully evaluate all potential models for public transport and, based on these, the recommendation by the EU Commission is:

> A combination of transport authority planning and control of public transport services on the one hand and competition between independent operators for the operation of public transport services on the other has the strongest merit.

It is to be noted that the Commission does not exclude publicly-owned operators from competition, if it is on the same conditions.

… and the implementation

The EU recommendation is thus clear as to the need for, and the direction of, a reform process. However, most Europeans have not yet introduced these reforms, but have kept their public monopolies. This is not so much out of a lack of conviction as for political and other reasons.

It should be remembered that it took several years for the UK to deregulate, and far-reaching structural changes may sometimes benefit from a little patience. (This is something that Western advisors occasionally appear to neglect when dealing with developing countries).

Progress of public transport reform in Europe

Regulated monopoly	Controlled competition	Deregulated market
Austria Belgium Germany Spain Luxembourg Greece Netherlands Italy Portugal	Sweden (route bidding) Denmark (route bidding) Finland (route bidding) Norway (route bidding) London (route bidding) France (management contract)	UK Baltic states

Table 5.1. Progress of public transport reform in Europe

Generally speaking, the major European countries have still to initiate the recommended reform programs for public transport, as they have maintained their public monopolies. Although the table above shows Western Europe, the same is true of the majority of new EU members from the former socialist Eastern Europe. (In the countries listed above as having predominantly regulated monopolies, there may be individual cities that have initiated reform.)

The forerunners of the controlled competition scheme in Europe are the Scandinavian countries and, to some extent, France. However, the difference is that in Scandinavia and London, routes or route packages are subject to bidding, while in France management contracts are being tendered.

The main deregulated market is still the UK. However, after their independence from the Soviet Union, the Baltic states (Estonia, Latvia and Lithuania) have promoted market economy reform and liberalization, including the public transport sector. As a result, the familiar pattern in developing countries has appeared in Baltic cities, now dominated by minibuses. (This is true of many Russian cities).

The case of Jamaica

Turbulent history

To reform a public transport system may be difficult enough in well-organized European countries, but it is much more difficult in a developing country. One of the examples of the dilemma of public transport reform is the city of Kingston, capital of Jamaica, which has a turbulent history.

1960s	British bus company
1970s	Nationalization
1980s	Privatization
1990s	Restructuring, merger
2000s	State-owned bus company

Table 5.2. History of public transport organization in Jamaica

In the 1960s, under British rule, a well-managed British bus company operated a much-appreciated service.(Older Jamaicans still remember it).

After independence, the company was nationalized and run as a public monopoly. Efficiency and service levels went down and the assets were eroded.

The Government followed the advice of World Bank advisors to privatize and deregulate the public transport sector in Kingston. The bus company was scrapped, and a minibus sector emerged. Due to fierce competition in the streets and a general lack of control, the situation became intolerable. There were riots in the streets of Kingston as people demanded buses.

In the 1990s Jamaica attempted to introduce a system of controlled competition from one day to the next, again on the advice of World Bank advisors. The attempt failed, since the concept of deregulation was deeply rooted and there was no longer any experience of running an organized bus company.

Jamaica has now gone back to the concept of a state-owned monopoly operator as the only way to introduce reform. At present (2003) the ambition is to build up an efficient bus company with the assistance of outside experts. The development in Jamaica can be described in the framework of the model given at the beginning.

<u>Moving North-East</u>
What happened in Jamaica in the 1980s was a step from the "Classical European Model" to a situation of uncontrolled competition and a resulting fragmentation of services. Deregulation was the watchword among many international advisors, and Jamaica followed. As many other countries have done, Kingston thus went from the South-West sector to the North-East.

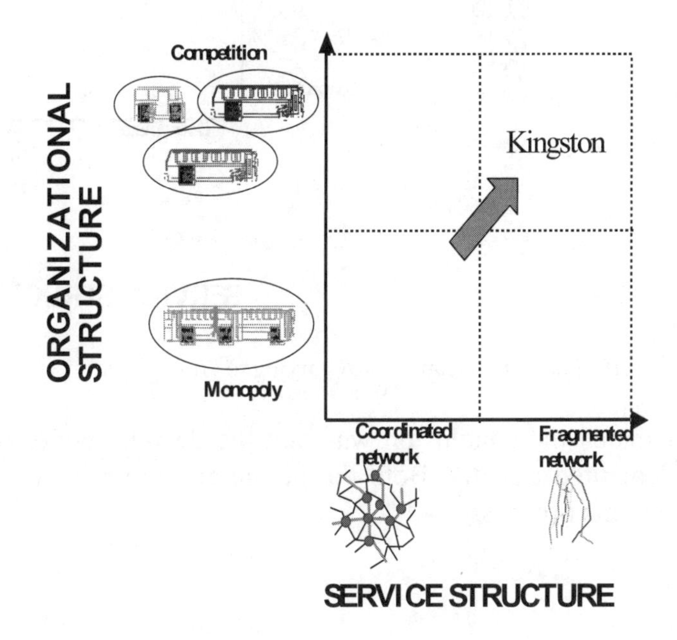

Figure 5.9. Public transport reform – "going North-East"

Trying to move North-West

In the 1990s it became increasingly clear to the international community that total deregulation might not be the best system. Kingston now attempted to restructure the system, and introduce controlled competition, and in so doing tried to go directly from the North-East sector to the North-West. However, this reform policy failed because of shortcomings in both the private and the public sectors.

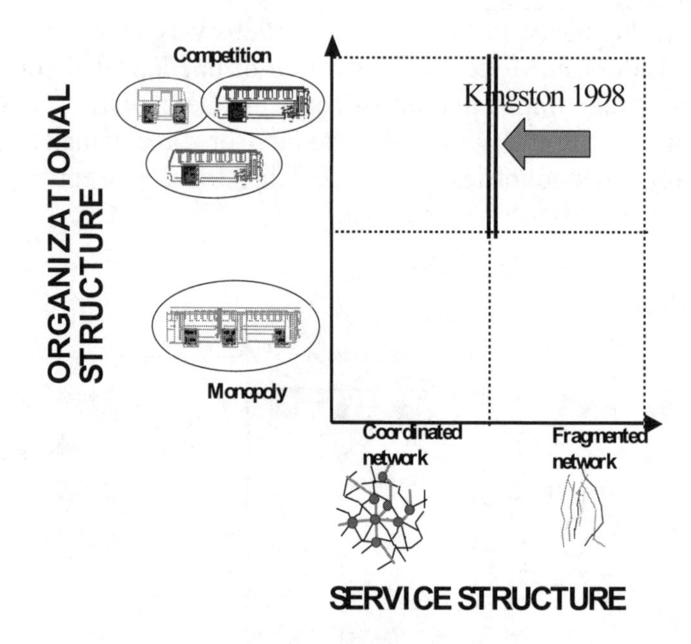

Figure 5.10. Public transport reform – trying to "re-regulate"

In Jamaica the conclusion was that the desired model could not be implemented directly. Both the public and the private sector play a role in this process.

Public sector	Private sector
Complex government organization	Too few actors
Weak institutions	Bus operation experience lost in Jamaica
Poor legal framework	Franchise holders investors, not operators
Lack of effective monitoring	Resistance to change
Unrealistic expectations	

Table 5.3. Shortcomings in both the private and the public sector

Moving back to South-West again

The city now decided to go back again from the North-East sector to the South-West; that is to reinstall the concept of a public sector owned bus company. The present challenge in Kingston is to make this work, and to develop an efficient company – something that will probably prove difficult.

If and when this is successfully achieved, Kingston might be able to follow the European example and introduce gradual reform. The development will be interesting to follow, but the outcome is far from certain, and it will, of course, take many years before the whole strategy can be evaluated.

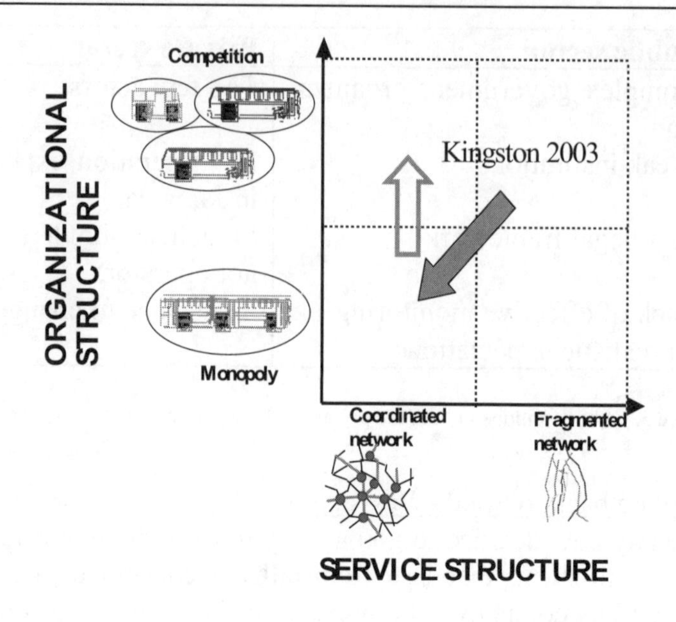

Figure 5.11. Public transport reform – going back to "South-West" again

Conclusions

The reform process in Europe is well in line with current international thinking. The World Bank, for example, now promotes this kind of "public/private partnership" in many areas, including public transport. Many countries recognize the advantages of the "controlled competition" concept, and would like to implement it.

It is, however, important to consider the difference in the preconditions of European countries and many developing countries. The dilemma with the "Classical European Model" was not that the service levels were particularly bad - although some state-operated systems may have been rigid and unresponsive to passenger demand. But efficiency in the monopolies was considered to be low because of many years of a protected and risk-free environment.

Europe	Developing country
- From monopoly situation - Increase number of operators - Split 1-2 monopolies - Maintain coordinated structure - Difficult, time consuming - Encouraging success stories when preconditions are right	- From fragmented situation - Reduce number of operators - Merge 1 000-10 000 actors - Create coordinated structure - Even more difficult - Few, if any, success stories when situation has got too far

Table 5.4. Public transport reform – different preconditions

The main issue, therefore, was to break up the old monopolies, something that can be difficult enough. (In e.g. St Petersburg, the three former monopoly operators are still practically intact in spite of over ten years of a new economic system).

In a deregulated developing city with hundreds - or thousands - of operators in a fragmented network, the problem is not to split but to *merge*. Once a city has allowed a situation to emerge in which there is a large number of operators in a loosely controlled and monitored system, then a reform in the direction of controlled competition may become very difficult. There are many examples of this. One is the attempt by the new non-apartheid government in South Africa to come to grips with a minibus sector which is practically in a state of war over profitable routes, and where frequent shoot-outs have become a serious problem.

It is perhaps time for international advisors and financial institutions such as the World Bank to ask themselves some questions that can serve as guidance in the future, e.g.:

- Have we been too dogmatic in our faith in market forces?
- Have we pushed cities into irreversible situations?
- Have we over-promoted the private sector, and neglected the role of the public sector?
- When is organized public transport feasible, and when is it not?
- What advice should we now give to cities that still have an option (China, India, Russia)?

Chapter 6. An Analysis of the Situation in Dar-es-Salaam in Tanzania from an Institutional Coordination Perspective

Kanyama, A. Carlsson-Kanyama, A. Lindén A-L. Lupala, J.

Introduction

The world's urban population is growing at a fast rate, posing a huge challenge to those responsible for the management of urban development and provision of services. In the African continent, urban population is growing faster than in any other continent, with a growth rate that reached 4.4% per annum between 1990 and 1996. The urban population is expected to more than double between 1990 and 2010, from 201 million to 468 million (Kanyama 1999). The continent is currently experiencing an average population growth of 4.7% per annum. In the 1980s the urban population growth rates for Kenya, Tanzania and Zimbabwe were 7.7%, 6.6% and 5.9% respectively (Mbara, 2002). It is predicted that by 2030, nearly 800 million people - approximately the population size of the entire continent today - will be living in urban areas (Mosha, 2001). In 2000 Cairo and Lagos were the only African cities whose population exceeded 10 million inhabitants. By 2020 the urban areas of Nairobi, Johannesburg, and Abidjan will also have reached or exceeded the 10 million mark, while 77 cities - six times more than today - will have more than 1 million people. There will be nearly 650 towns - close to four times the current figure - with more than 100,000 inhabitants. The built-up areas will be 12 times as dense, and the average distance between towns four times shorter (Ibid, 2001). This development is already manifesting itself in Africa's sprawling urban structures, which make the journey to work excessively long and costly for some of the very poor. In addition, the increase in population has pushed up the demand for public transport and led to the emergence of transport services that are of very poor quality. In most African cities, normal bus services have all but disappeared and been replaced by minibus paratransit vehicles that go by such colourful names as Danfos – Nigeria, Matatus – Kenya, Tro-tros – Ghana, Dala-dala – Tanzania, Combis – South Africa, and Car Rapides – Senegal (White & Hook, 2002).

The public transport services are being provided by such buses no longer meet the expectations of the people owing to their negative environmental and social impacts. These buses constitute a system of public transport that has such characteristics as: (i) unmanaged routes and trip frequency, which lead to long queues and road congestion; (ii) unreliability of services; (iii) excessive energy consumption, as vehicles often accelerate/decelerate while at low speeds or keep their engines running when the buses are stationary; (iv) excessive gaseous emissions due to poor maintenance of vehicles, with resulting increases in air pollution; and (vi) increased traffic accidents, as unsafe and reconditioned vehicles are used for many years and some of them are not roadworthy. As a result, travelling becomes uncomfortable to people and affects their quality of life. These are the visible manifestation of failures in urban transportation planning today.

As population continues to increase in Africa's urban areas, it seems that nothing can stop the breakneck expansion of these cities and the deterioration of public transportation and the environment unless drastic action is taken soon. Most governments in Africa have taken the view that they need to control and manage the process of urban growth. Without a doubt, many governments have adopted specific policies to this end, but few appear to have been successful. Certainly, there are a number of factors that may constrain the achievement of sustainable urban public transportation, such as lack of capital and know-how. However, as this paper shows, one of the main reasons preventing the creation of a viable plan for a sustainable transport system in African cities may be the lack of co-ordination among planning institutions and other stakeholders who have a role to play in planning for better urban transportation. Dar-es-Salaam city forms the case study for this paper, which focuses on the inefficiency of public transportation viewed from the perspectives of city growth and institution accountability. The methodology for the study included literature reviews on emerging models of sustainable urban development, reviews of official documents that have a bearing on public transport in Dar-es-Salaam, and interviews with officials in public institutions and agencies, as well as with residents in selected neighbourhoods.

Fast population and spatial growth - a challenge for efficient public transport in Dar-es-salaam

Tanzania, the country selected for this case study, is a low income country according to the classification system adopted by the World Bank, meaning that the per capita income is US$ 735 or less per year, although today Tanzania is classified as a less indebted country (World Bank, 2004). Tanzania is situated in East Africa, bordering the Indian Ocean in the east, Kenya and Uganda in the north, Rwanda and Burundi in the north-west, the Democratic Republic of Congo in the west, Malawi and Zambia in the south-west and Mozambique in the south.

Dar-es-Salaam, the capital and largest city of Tanzania, dates back to the German colonial era of the late 1800s. In 1891 the extent of the built-up area of Dar-es-salaam city was limited to only 122 hectares, with a population of about 4,000 inhabitants (Kombe, 1995). This figure had increased to 463 hectares in 1945. During the 1940s and 1950s the Dar-es-Salaam city boundaries averaged less than 5 kilometres from the sea front or the existing town centre. Between 1945 and 1963 the built-up area of the city extended to 3081 hectares, while by the year 2002 the population of Dar-es-Salaam stood at 2,497,940 within a built-up land area of 57,211 hectares (Lupala, 2002). The city's spatial expansion has been occurring at an average rate of 7.2% per year. During the post-independence period (i.e., after 1961) a rapid population influx from the countryside and many individuals building their own houses has resulted in a horizontal expansion of the city, predominantly along its radial network. By 1978 the built-up parts of the city extended 14 kilometres along one main road (Pugu) and about 12 kilometres along other two main roads from the city centre (Bagamoyo and Morogoro). By 1992 the extent of the built-up area predominantly remained within a 12 to 16 kilometre radius, while in 2002 the built-up area extended to a 32 kilometre radius (ibid). Most of the post-independence city expansion has taken place in an informal manner, with housing structures that are predominantly low-rise and single-unit types surrounded by gardens used for cultivation and some animal husbandry. This sprawling low-density city lacks basic social facilities and infrastructure services and its inhabitants turn to the inner part of the city for employment opportunities and major services. This has increased the demand for travel into and out of the city centre.

Public (Bus) transport in Dar-es-Salaam

The development of public transport in Dar-es-Salaam

Public (bus) transport in Dar es Salaam dates back to the British colonial era when in 1949 a privately owned British company known as the Dar es Salaam Motor Transport Company (DMT) was started to provide bus services in the city. The good quality of public transport services offered by DMT continued relatively well until the mid-1970s. According to bus commuters of that time, buses adhered to timetables and delays were minor.

In 1970, DMT was nationalised, and in 1974 it was renamed 'Usafiri Dar-es-Salaam' (UDA), meaning literally 'Public Transport in Dar-es-Salaam'. The nationalisation of DMT was in line with the then socialist ideology that the 'commanding heights' of the economy ought to be under the control of the state. UDA was thus owned jointly by Dar es Salaam City Council with 51% of the shares, and the National Transport Company (a government agency) holding 49% of the shares (Kombe et al. 2003). As the sole provider of bus services in Dar-es-salaam, UDA operated fairly satisfactorily immediately after it acquired the assets of DMT. It inherited good quality buses that were comfortable and well-suited to the city's public transportation. It also carried forward DMT's basic transport planning skills among its staff, namely planning for bus route networks, number of routes, route length, bus terminals and principal bus stop locations. However, by and large, UDA operated under the auspices of the government and therefore the fare levels it set had to get cabinet approval. Fare levels were regulated according to what the government thought the majority of the people could afford to pay, with no investigation or consideration of actual operating costs. The fare levels sanctioned by the government were too low to cover operating costs, and the government could not cover the financial gap. The financial deficits that occurred as a result of this situation had a devastating effect on the efficiency of UDA in its delivery of services (Kanyama et al. 2004). This problematic situation occurred at the same time as a dramatic growth in travel demand during post-independence. Dar-es-salaam's population grew rapidly (Figure 6.1).

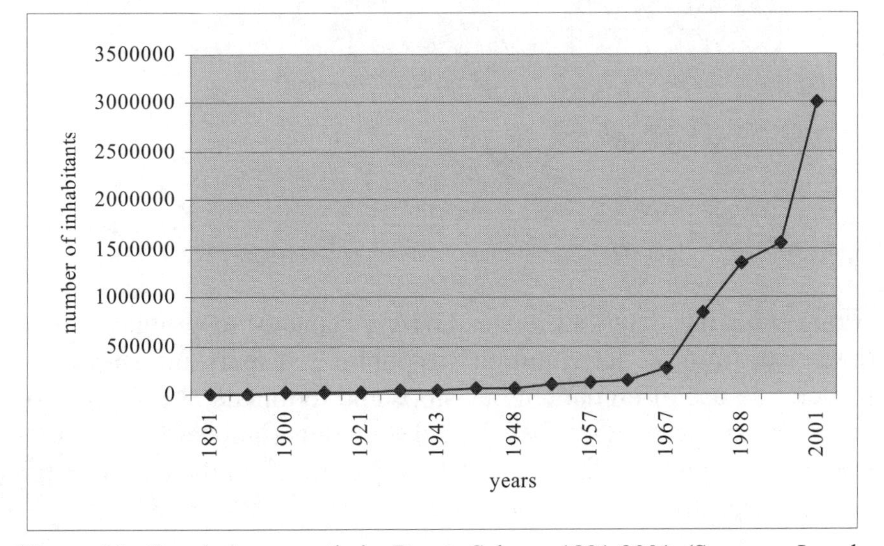

Figure 6.1: Population growth in Dar-es-Salaam 1891-2001 (Source: Lupala, 2003).

The number of UDA buses declined from 1975 (Figure 6.2). The bus fleet included standard single-decker buses with a carrying capacity of 90 passengers and minibuses with a carrying capacity of 30-50 passengers. (Source: URT, Budget Speech for Year 1983/84, cit. Kombe et al. 2003.

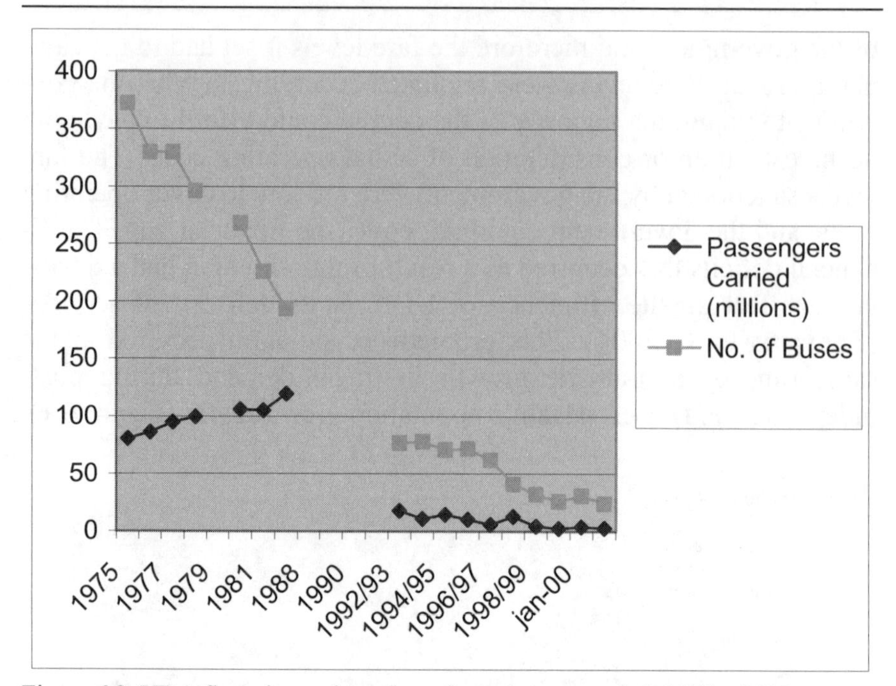

Figure 6.2. UDA fleet size and number of passengers carried, 1975 – 2000.

Along with the deterioration in UDA's capacity to provide buses, there was also a deterioration in public transport infrastructure, namely the use of unmarked bus stops and terminals, the absence of bus bays, shelters, posts, benches, destination signboards and timetables. On the whole, the main factors that impaired the public transport services offered by UDA were:

- Lack of adequate finance (from the main financier, the Tanzanian government) to purchase buses and spare parts to extend the bus fleet in order to meet public travel demand
- Lack of qualified technicians, engineers and transport planners to carry out maintenance and scheduling of vehicles
- Poor road conditions characterised by extensive potholes that inflicted mechanical damage on the buses. As a result, UDA services became further constrained by an increasing number of unserviceable buses
- Low fare rates that did not reflect the real market price for travelling

Emergence of privately owned dala-dala buses

Illegal private transport operators emerged in the late 1970s and early 1980s due to the gross failure of UDA to offer adequate transport services to meet public travel demand. These illegal operators charged a fare of 5 shillings instead of the 1-1.5 shillings charged by UDA, yet the demand for the illegal operators was high, a clear testimony to the enormous gap between the bus services supplied by UDA and the public travel demand. The services of the illegal operators were popularly named 'dala-dala' because of the then exchange rate of five Tz. shillings to one US dollar. Since then, dala-dala has become the term used to refer to all privately owned buses providing public transport services in the city (Kanyama et al. 2004). When dala-dalas first emerged, the government authorities made numerous efforts to prohibit their operation but such efforts were unsuccessful and the dala-dala buses continued to increase to meet existing demand.

Until the beginning of the 1980s conventional buses were the dominant public transport mode used in most African cities. There was an ambition to subsidize them in order to keep the fares at an affordable level. However, in Africa generally government-supported public transport initiatives were greatly affected by the Economic Reform Programmes (ERP) which were introduced in the 1980s by many governments. Economic Reform Programmes in essence were intended to create an environment that encouraged competition by providing entrepreneurs with the freedom necessary to respond to market opportunities and not be constrained by economic controls. The deregulated regime had no place for subsidies, and in most cities conventional buses have disappeared, as is the case in Dar-es-Salaam. It is now very common for public transport to be provided by private operators using smaller vehicles (Mbara, 2002).

In 1983 the government of Tanzania issued a directive officially to allow the operation of dala-dalas in order to solve the city's transport problem. This was in part to comply with the Economic Reform Programme (ERP) and also resulted from the persistent public transport problem in Dar-es-Salaam. The official recognition of dala-

dalas as a mode of public transportation compelled the government to set up a system for organising the operations of the dala-dala buses. Accordingly, the Central Transport Licensing Authority (CTLA), a department within the Ministry of Communication and Transport, was given the responsibility for issuing licenses to vehicles operating in Dar-es-Salaam. All the operators had to apply to the CTLA for licenses and route permits. The CTLA did not issue licences based on the right to ply specific routes, but instead issued permits and road licences valid on all Dar-es-Salaam roads. For that reason, competition and chaos among dala-dala operators became common as buses could operate on any route which they thought had more passengers.

In view of the chaotic operations of dala-dalas that became evident in Dar-es-Salaam, the Regional Commissioner initiated a takeover of management of dala-dala services from the CTLA. In 1999, the Regional Commissioner formed an agency - the Dar-es-Salaam Region Transport Licensing Authority (DRTLA) - to license commuter buses within the Dar-es-Salaam region.[2]

Today the process followed by a prospective bus operator in Dar-es-Salaam starts with the submission of an application to the DRTLA. To obtain a license and route permit the operator must submit a vehicle registration card, business licence tax clearance certificate and vehicle insurance documents along with his initial application. Thereafter, the vehicle must be presented to the traffic police for a roadworthiness inspection. Once the vehicle inspection certificate has been obtained, it must be presented to the DRTLA, which considers the application and grants the bus-operating licence. The licence is usually issued for 6 months or one year at a time (depending on the financial status of the applicant). Vehicles are expected to operate on specific roads. Each route is allocated a certain number of vehicles, and each vehicle is given a colour for that specific route.

However, the DRTLA lacks the professional competence to carry out proper bus route planning. Unlike the UDA or DMT, the

[2] Interview with executive secretary – DRTLA

DRTLA had neither traffic inspectors nor traffic planners among its staff. Instead, it recruited field assistants employed on a temporary basis to carry out transport planning.[3] The DRTLA has a shortage of finance to recruit and pay field assistants, and for that reason it is hard for the DRTLA to retain even low-skilled staff. The secretary of the DRTLA (in our interview) concedes that today they assign bus routes merely based on intuition and simple methods including: i) assessments made by traffic counts on different routes; (ii) simple on-site observations of concentrations of people in different areas in the city; and iii) hasty and generalised assessments of income distribution according to the quality of built-up areas.

Furthermore, the DRTLA acknowledges that problems constraining the management of (dala-dala) public transport in Dar-es-Salaam include: i) a lack of inspectors who could ensure that operators adhered to the route to which they were allocated; ii) only 30% of dala-dala vehicles are roadworthy despite the existence of regulations banning such vehicles from the roads; iii) the high accident rates due to poor driving skills; iv) the overcrowding of passengers in buses and congestion of vehicles on the roads, which leads to delays in arriving at destinations.

A recent development concerning privately-owned vehicles for public transportation in Dar-es-Salaam is that in 2003 the DRTLA announced a strategic plan aimed at banning small dala-dalas, commonly known as 'vipanya' (mice), from operating on city centre roads so as to reduce congestion in Dar-es-Salaam. In a statement, the Chairman of the Dar-es-Salaam Regional Transport Licensing Authority (DRTLA), said: "*We have not and will not change our minds - the plan to phase out minibuses (dala-dala) from the city centre roads is there*." The Chairman reinforced this by referring to the inefficient public transport system in Dar-es-Salaam in terms of increased road accidents, traffic congestion caused by vehicles and pushcarts, etc. (*Guardian*, 2004.01.15). Under the proposed plan, minibuses will be required to pick up and drop off commuters in the peripheral areas of the city centre such as Kariakoo. Big buses,

[3] Interview with executive secretary – DRTLA

dubbed 'city buses', will then transport the passengers to city centre areas like the main Post Office (Posta) and the ferry (Kivukoni). "*We are slowly phasing out minibuses, and I hope they will be out of the central business district (city centre) roads*," said the Chairman of the DRLTA. He went on to explain that the implementation of the plan would be gradual and there was no way that it could be effected abruptly - but declined to give a precise deadline: "*It is not possible to impose a ban on minibuses in a day. We will cause numerous problems to city residents who solely depend on these services*."(ibid.). It is of interest to note that the plan contemplated by the DRTLA is not clearly coordinated with a plan to modernise public transport by the help of Rapid Bus Transit (RBT), which is being initiated by the Dar-es-Salaam City Council (see below).

City Council recognition of its role in managing public transport in Dar-es-Salaam

The City Council sees itself as the rightful agency to manage public transport in Dar-es-Salaam according to the 1997 Transport and Licensing Act which was amended in 1999. On the other hand, the City Council has no department equipped with competent transport planners and engineers who can adequately plan and manage public transport in Dar-es-Salaam.[4] This is why the DRTLA, despite being inadequately professionally equipped to manage public transport in the city, is assured of the role for some time to come. The City Council also recognises its legal role in planning and managing public transport. In light of this, and as a result of chronic public transport problems, the mayor of Dar-es-Salaam announced that the city will begin to develop a new city-wide mobility plan for public transport. Speaking at a workshop organised by the city council on May 13, 2003, the mayor said that the proposal will include plans for a world-class Bus Rapid Transit system (ITDP, 2003). According to a City Council planner, the planned Bus Rapid Transit (BRT) would either replace the existing poor and chaotic public city transport vehicles (dala-dalas) or reorganise them to operate more efficiently alongside BRT. The BRT vision is to embark on a modern public transport system at a reasonable cost to the users with quality and

[4] Interview with the City Council Planner in mid-2003

high-capacity buses that meet international service standards, reduce travelling time and are environmentally friendly. The target is to make the BRT project operational by 2005 (*Guardian*, 2003.10.11). Members of the Dar-es-Salaam City Council travelled to Colombia in 2002 with some representatives of the Dar-es-Salaam Bus Owners Association (DABOA) on a study tour to see how BRT is managed. The City Council has thus identified the Colombian capital, Bogota, as a place from which to learn how BRT works.

What is Bus Rapid Transit? This is a new paradigm in delivering bus services which is being developed in a number of countries, particularly in Latin America, and it shows promise for revolutionising bus systems around the world (IEA, 2002). As argued by the International Energy Agency (IEA), separating buses out of the general traffic by assigning specific lanes, increasing their average speed, improving their reliability and convenience, and increasing system capacities can ensure high usage levels and increase the profitability of the Bus Transit system (IEA, 2002). Upgrading the performance of bus services to meet the objectives of Bus Rapid Transit will require policies that give priority to bus operations and provide for investment in crucial system components: infrastructure that separates bus operations from general-purpose traffic; facilities that provide for increased comfort and system visibility; and technology that provides for faster and more reliable operations (FTA, www.fta). New guidance, information and fare technologies offer an expanded range of possibilities for operating bus systems that have the potential for producing marked improvements in performance, surpassing previous standards and changing public perceptions of bus services (ibid). However, Bus Rapid Transit provides maximum benefit when developed in close coordination with land-use policies and community development plans. These operations will require improved land-use options that provide for compact, pedestrian-friendly and environmentally-sensitive development patterns that can sustain the development of Bus Rapid Transit.

Clearly, implementing Bus Rapid Transit poses a number of challenges in Dar-es-Salaam. These include the need for adequate lanes on city streets to provide separate rights-of-way for buses yet maintaining the quality of general-purpose traffic flow. These challenges require detailed analysis in the context of specific local applications to identify appropriate solutions and to determine where Bus Rapid Transit can have the greatest benefit.

Under its proposed scheme, the Bus Rapid Transit (BRT) programme, the City Council has earmarked some major roads in Dar-es-Salaam for expansion in order to improve the city's public transportation. Roads identified in the initial phases of the BRT programme include Nyerere, Mandela, Morogoro and Ali Hassan Mwinyi. The intention of the City Council is to expand these roads in order to provide room for another lane to be used by passenger buses that will serve the public under the BRT programme. No other vehicles will be allowed to use these additional lanes set aside for BRT. According to the statement from the City Council, the buses to be used in the BRT programme would be owned and operated by private firms (*Guardian*, 2004.01.20). In addition, the City Council plans to conduct a number of surveys in 2004 to collect standard data for use during the planning and other stages of execution of the project. These will include an Origin and Destination (OD) survey to determine the number of commuters from each part of the city and their travelling patterns in order to allocate a sufficient number of buses at appropriate times (ibid).

Constraints on improving public transport in Dar-es-Salaam today

Clearly, the present chaotic public transportation situation in Dar-es-Salaam reflects the on-going and rapid changes transforming a less-industrialised country into a market economy during a period of rapid urbanisation. The result of this partly unplanned, unmanaged and under-financed process is substantial discomfort for that large proportion of the population that depend on public transport for their everyday activities, including schoolchildren and the elderly.

Institutions with a stake in transport management and planning in the city are today deeply concerned about the situation and have proposed various plans for change, including the introduction of Bus

Rapid Transport. In this endeavour it is important to recognise and deal with the institutional constraints on implementing the Bus Rapid Transport system and improving public transport in general in Dar-es-Salaam. According to our informants and our analysis, such constraints can be expected on three fronts, namely: (i) from the mismanagement of the *physical form* of the built-up area; (ii) from lack of *coordination* within the City Council and the Regional Authorities (Dar-es-Salaam Regional Traffic Police Department and Dar-es-Salaam Region Transport Licensing Authority); and (iii) from *lack of institutional coordination among government sectors.* In the following section we examine these three constraints briefly and draw some conclusions.

Inability of institutions to manage the physical form and the built up area in Dar-es-Salaam

Narrow roads dominate the inner part of the city because they were planned during the colonial period when the traffic volume was smaller than it is today. It will be hard to expand the roads to increase their capacity by demolishing buildings constructed close to the roads. Basically the government does not have sufficient financial resources to compensate for properties that would be demolished. However, the narrowness of roads in Dar-es-Salaam is not only attributed to the physical characteristics of the older fabric of the city. To a greater extent, the contemporary practice of granting permits to build closer to the roads reduces the capacity of the roads by taking over space that is meant for parking, walking and cycling.[5] Such discrepancies in planning and management of the city arise partly due to the insensitivity of practitioners in heeding the guidelines set for the development of the city. Another major observable land use phenomenon in Dar-es-Salaam is what Kironde (1995) calls 'spontaneity', which means that residential land and open space is being converted for commercial use, particularly along major thoroughfares, around bus terminals and main road junctions, without government sanction. In most cases, these conversions or constructions have neither planning consent nor building permits, although the businesses being carried on usually have trading licences. Yet

[5] Interview with a City planner

neither the central government nor the Dar-es-Salaam City Council has deemed it necessary to take any action (Kironde, 1995). Similarly, it is common to see street traders occupy space designed for pedestrian and vehicular traffic in Dar-es-Salaam. This effectively reduces the capacity of the road, causing congestion and thus affecting traffic flow and compromising safety. So far, the lack of alternative spaces for both street traders and parking places in many areas of Dar-es-Salaam has complicated the work of the traffic police in enforcing the law against the violations of traffic regulations.[6]

Our interviews with City Council officials showed that there is a general acceptance in the City Council Authority that future land use planning in the city and its implementation should be carried out in parallel with transport planning, yet there is a continuing disregard for integrated land use development and transport planning. It is common to find that the construction of residences in Dar-es-Salaam precedes transport planning. In the same way, the City Council has neither made efforts to determine the capacity of its roads nor investigated transport demand for different areas of the city. Furthermore, it is common to encounter permits that contravene the master plan - a permit may be granted to develop a certain area contrary to its original use without regard for how much traffic it can generate. As an example, an open space in the centre of the city, part of the Gymkhana grounds, was carved off in 1993 and allocated to a private firm for the construction of a Sheraton Hotel. There were various protests against the scheme from sports lovers, environmentalists and a number of the city residents who respect open spaces, but to no avail. As argued by Kironde (1995), the decision to locate the hotel on an area of the Gymkhana recreation ground was political, and was made at a higher level than the Ministry of Lands. In light of this, a conclusive statement by a City Council planner during our interview was that "the quest for sustainable public transportation in Dar-es-Salaam is restrained by the way urban development is mismanaged in the City".

[6] Source: Interview with Dar-es-salaam Region Traffic Police authority

Lack of a coordination mechanism at the City Council and the Regional Authorities in Dar-es-Salaam

As has been pointed out earlier, the City Council lacks adequately trained transport planners and engineers who can manage public transportation in the city, and there is no organized department specialised in transport planning. The City Council sees the coordination between itself and the sub-areas of Dar-es-Salaam as inadequate. For instance, whenever attempts are made to convene meetings for the City Council's working groups to discuss issues concerning sustainable development, it is common that attendance is poor among municipal engineers and planners from the sub-areas of Dar-es-Salaam. According to our interviews at the City Council, inadequate coordination in the sector of public transport between different units arises due to a lack of clear policy statements, as well as a lack of accountability on the part of respective officers.

Regarding coordination among Dar-es-Salaam Authorities in general, an official with the Dar-es-Salaam Region Transport Licensing Authority (DRTLA) believes that poor coordination involves all the institutions that have a role in the operation of public transportation in Dar-es-Salaam. These include the DRTLA itself, the City Council, the Dar-es-Salaam Region traffic police department, sub-area councils etc. In view of that, since there is no well-organized system of coordination between these institutions, overlapping areas of responsibility are common, and since there are no mechanisms for managing overlaps, conflicts are common in the execution of roles.

Inadequate institutional coordination among national government sectors

The Ministry of Communication and Transport is responsible for issuing a national transport policy, which includes policies for safer, more effective and environmentally friendly transport systems. The Ministry is also responsible for reviewing policies and for making recommendations on new policies. According to the present National Transport Policy, NTP (2003), the Ministry is also supposed to develop a safe, reliable, effective, efficient and fully integrated transport infrastructure and operations which will best meet the needs of travel and transport. The Ministry of Transport and Communication acknowledges that the transport sector in Tanzania is characterised by low quality services for various reasons including:

- The existence of a great outstanding need for infrastructure maintenance and rehabilitation work
- Inadequate institutional arrangements, laws, regulations and procedures which are inconsistent or incompatible with each other for the creation of a conducive climate for investment and hence growth in the sector
- Inadequate capacity caused by low level of investment in resources
- Low levels of safety enforcement, environmental sustainability, and gender issues

According to the National Transport Policy, persistent weaknesses in the development and management of the transport sector are to be found in the Ministries responsible, which are:

- Communication and Transport (Road Transport Licensing)
- Works (axle-loads control, safety control)
- Home Affairs (traffic law and regulations enforcement)
- Finance (motor vehicle registration, road tolls)
- Regional Administration (regional transport licensing)
- Vice-President (environment)
- Planning Commission (key transport utilities)
- Trade and Industry (vehicle licensing)

According to the NTP, major weaknesses inherent in the above-mentioned institutions in connection with the transport sector include poor coordination, poor governance (corruption) and poor enforcement. On the other hand, whenever Ministries take steps for coordination, they turn out to be erratic and lack the basis for continuity. They are often characterised by:

- Lack of coherent policy guidance on the part of those concerned with the planning and development of the transport sector, leading to disjointed plans and programmes
- Inadequately formalised coordination and consultation among principal actors
- Shortage of adequately trained and experienced personnel in the transport planning departments and units
- Non-application of scientifically based planning methodologies fitting the Tanzanian environment, coupled with non-existent data systems
- Inadequate infrastructure and facilities to cater for non-motorised transport such as carts and bicycles and also simple motorised equipment such as motor cycles and similar intermediate technology facilities
- Low managerial capacity in public enterprises; underfunding of such enterprises and the absence of meaningful competition
- Lack of regulatory regimes that are adequately equipped to enhance competition, fair operational practices and complementarity of services
- Insufficient dialogue between the public and private sector due to poorly developed service users or consumer associations

Although the Ministry of Communication and Transport issues policies for transportation and has a coordinatory role aimed at ensuring that the operation of transportation becomes sustainable, a respondent in the interview (held in 2003) observed that he had not seen the Ministry carry out a concrete or systematic research or programme to tackle the problems of public transport in Dar-es-Salaam. Other institutions crucial for traffic management such as the Police Traffic department in Dar-es-Salaam, expressed frustration at their lack of involvement: "When the Ministry of Works embarks on building Dar-es-Salaam roads, the Traffic Police Department is not consulted, although its staff have everyday experience of traffic management which could be accommodated during the design of roads and in the provision of associated infrastructure"

As pointed out above, the Ministry of Communication and Transport is responsible for issuing a national transport policy, which includes policies for a safer, more effective and environmentally friendly transport system. In addition, the Ministry is expected to spearhead the development of safe, reliable, effective, efficient and fully integrated transport infrastructure and operations which will best meet the needs of travel and transport. It is, therefore, evident that the Ministry ought to shoulder the coordination role to promote sustainable public transport in urban areas. Yet, what we have seen above in this paper is that the coordination among institutions, government sectors and agencies to promote sustainable public transportation is dismal in Dar-es-Salaam.

Conclusions

Dar-es-Salaam city is experiencing unprecedented growth, which is placing great pressure on services. Public transport, which is central to development, is one of the services that has to be provided in an expanding city. Thus, city administrators are faced with immense challenges in developing a sustainable transport system that is responsive to changing demands. However, a lack of clear policy guidelines for public transport and a lack of a clearly identified coordinating body has resulted in poor coordination, which makes planning and managing public transport in Dar-es-Salaam more difficult than it would be even considering the limited capacity to generate tax revenues. For instance, both Dar-es-Salaam Region Transport Licensing Authority and the City Council struggle to manage public transport in Dar-es-Salaam without any clear mechanism of coordination among themselves. Lack of coordination extends also to the Ministries and agencies that have a stake in public transportation in Dar-es-Salaam.

Effective public transport and traffic planning in Dar-es-Salaam should involve new ways of designing and implementing policy as well as new policy objectives. These new ways of working will require clear action by national government to put the right conditions in place allowing local authorities to do their work effectively. A clear guidance policy framework must be backed by adequate human resource requirements. The City Council, together with its sub-area Councils, needs to develop its own transport sector policies that are consistent with the national transport policy. Such a policy needs to articulate the objectives of urban transport in relation to economic growth, infrastructure maintenance, provision of an affordable and efficient public transport system particularly for the urban poor, minimising transport resource costs by making more effective use of existing facilities, and minimising the impact of transport on the environment. Developing possibilities for non-motorized transport is crucial. Awareness campaigns are imperative to inform the public about the seriousness of environmental and social problems arising from inefficient transportation. This is meant to motivate them to participate in schemes that can lead to efficient public transportation. The development of sustainable transport in Dar-es-Salaam must be accompanied by an environment that is conducive to development. A transparent and symbiotic partnership between central and local governments, the private sector and civic societies has to exist. These stakeholders have to share the common goals of developing an economically and environmentally sustainable transport system.

The analysis presented in this paper is an extract from publications of a research project with the title: '*Institution and Technical Constraints in Planning for a Sustainable Public Transportation: A Case Study of Dar-es-Salaam, Tanzania*' carried out from 2003 to 2004 and financed by Volvo Research and Educational Foundations. The title of the latest report is *Public transport in Dar-es-salaam, Tanzania: institutional challenges and opportunities for a sustainable transport system* (2004).

References

Federal Transit Administration (FTA) :
http://www.fta.dot.gov/brt/issues/pt5.html

Guardian, 2004.01.20 :
http://www.ipp.co.tz/ipp/guardian/2004/01/20/4147.html).

Guardian, 2004. 05.20 :
http://www.ippmedia.com/ipp/guardian/2004/05/20/11196.html).

Guardian 2004.05.18:
http://www.ippmedia.com/ipp/guardian/2004/05/18/11074.html).
Guardian, 2004.01.15:
http://www.ipp.co.tz/ipp/guardian/2004/01/15/3875.html
Guardian, 2003.10.11
http://www.ippmedia.com/guardian/2003/10/11/guardian6.asp.

IEA (International Energy Agency), 2002. *Bus System for the Future. Archiving Sustainable Transport Worldwide.* Paris. OECD

ITDP, 2003:
http://www.itdp.org/STe/ste8/index.html#dar_es_salaam

Kanyama, A. Carlsson-Kanyama, A. Lindén A-L. Lupala, J. 2004. Draft Report. *Institutional and Technical Constraints in Planning for a Sustainable Public Transportation: A Case Study of Dar-es-Salaam, Tanzania.*

Kanyama A. 1999. *Strategic town planning for environmentally friendly transportation: A comparative study of Dar-es-Salaam and Lusaka*. A research proposal presented to Sida -- SAREC.

Kironde, J. 1995. *The Evolution of the Land Use Structure of Dar-es-Salaam, 1890-1990: A Study in the Effects of Land Policy*. Nairobi. University of Nairobi.

Kombe W.J. 1995. Formal and informal land management in Tanzania, the case study of Dar-es-Salaam. In: *SPRING Research Series No. 13,* Dortmund.

Kombe W., Kyessi, G.A., Lupala J. and Mgonja E. 2003. *Partnerships to Improve Access and Quality of Public Transport: A Case of Dar es Salaam, Tanzania*. Water, Engineering and Development Centre (WEDC), Loughborough University, UK.

Lupala, J. 2003. The Spatial Dimension of Urbanisation in Least Industrial Countries: Analysis of spatial growth of Dar-es-Salaam city, Tanzania. A paper presented at the International conference on sustainable urbanisation strategies. Weihai, China; 3-5 November, 2003

Lupala, J. 2002. *Urban Types in Rapidly Urbanising Cities: Analysis of Formal and Informal Settlements in Dar-es-Salaam, Tanzania.* Stockholm. Royal Institute of Technology

Mbara T.C. 2002. How have African Cities managed the sector? What are the possible options? A Paper presented at the Urban and City Management Course for AfricaUMI, Kampala, Uganda, 4-8 March 2002:
http://www.worldbank.org/wbi/urban/docs/TRANSPORT-%20T%20MBARA.pdf

Mosha, A.C. 2001. Africa's future: Africa's cities must be transformed if the continent is to keep pace with the rest of the developing world. In: *Forum for Applied Research and Public Policy,* Summer 2001.

National Bureau of Statistics:
http://www.tanzania.go.tz/statisticsf.html

United Republic of Tanzania. 2003. *National Transport Policy*. Dar-es-Salaam. Ministry of Communication and Transport

White, P. S. and Hook, W. 2002. Africa's public transit renaissance. In: *Sustainable Transportation,* No. 14. ITDP.

World Bank. 1994. www.worldbank.org

Chapter 7. Towards Sustainable Urban Transport in China: The Role of Bus Rapid Transit Systems

Liya Liu

Introduction

The urban proportion of China's population is forecast to increase from 30.2% (319 million) in 1995 to 49.1% (711.7 million) in 2020. During the same period car ownership in cities is forecast to increase by 1,000%, i.e. ten times more. Urban traffic congestion is becoming rapidly worse. It is already a major constraint on urban productivity and sustainability and it is impeding improvement in the quality of life. Most urbanised cities have experienced serious traffic congestion. In some large cities, the duration of congestion, the size of congested areas and their number are on the rise. Vehicular air and noise pollution is increasing, contributing significantly to greenhouse gases (GHG) and global warming and posing a serious health threat to urban populations. Urban transport energy consumption has been growing steadily. Shares of traditional NMT (non motorised transport) modes, especially bicycle transport, are decreasing in response to both longer travel distances (brought about by an unprecedented urbanisation and urban sprawl pattern) and problems inherent in sharing road space with more cars. Much of the erstwhile NMT traffic is presently shifting to public transport modes, predominantly the bus. However, public transport services are poor in many cities. The development of public transport is often hindered by a lack of capacity, slow operating speeds, outdated equipment and outdated management practices. Urban commuters have been experiencing longer traveling times by using the bus.

Some cities have been expanding their roadway infrastructures assertively, the aim being to develop car-dominant transportation systems. Large metropolitan areas such as Shanghai, Beijing, Tianjin,

Chongqing, Guangzhou, Wuhan and others have been developing both the metro systems and expanding the roadway infrastructure. Even though some Chinese cities have developed isolated bus priority schemes, the concept of the Bus Rapid Transit system (BRT) has not been recognized as a cost-effective and sustainable public transportation mode.

The Institute of Comprehensive Transport (ICT) at the National Development and Reform Commission (NDRC), with the co-operation of international agencies and foreign private foundations, has been spearheading BRT development in China since 2000. This effort has significantly increased the knowledge of BRT systems on the part of decision-makers and technical communities in China. With the aim of piloting BRT projects in selected cites, BRT will become a new and affordable public transport mode for many Chinese cities, such as Beijing, Shanghai, and further promote sustainable urban transport development economically, socially and environmentally elsewhere in China.

Urban and Economic Development in China

China is a developing country with the largest population in the world. In the past 20 years, China's economy has experienced consistent and rapid growth. The gross domestic product reached an annual rate of growth of 9.4 per cent (on average) between 1978 and 2003. The GDP of major cities currently accounts for 61% of the total GDP. The urban population accounts for about 40% of the population. The Chinese government has set a national strategy goal of quadrupling the GDP by 2020, maintaining an average annual economic growth rate of 7.2% and aiming to build up a generally affluent society. According to international experts the first two decades of this century will be a key period for bringing about industrialisation in China, through which significant changes can take place in the nation's economic structure and urbanisation level and the population's consumption structure.

Astonishing Economic Growth

Since the new economic reform policy in 1978 involving 'opening up to the outside world', China's economy has, as mentioned above, been growing at an annual average rate of 9.4%. According to preliminary estimates and evaluations, China's GDP in 2003 increased to US$ 1.4 trillion from US$ 147.3 billion in 1978 and up 9.1% on the previous year at comparable prices. The per capita GDP has reached US$ 1,090, representing the highest growth since the Southeast Asian financial crisis in 1997. It was a hard-won success after the outbreak of the SARS epidemic and other frequent natural disasters. The World Development Indicator of the World Bank shows that China's GDP in 2003 ranked seventh in the world and accounted for 3.88% of the world's gross economic volume.

Expanding Urbanisation

As a consequence of Chinese economic growth and the change in its structure in recent years, China's economy has entered a stage of accelerated urbanisation, the dominant orientation being the chemical industry and other heavy industries. In order to realise the goal of quadrupling GDP, the level of urbanisation must rise, with the urban population increasing by 70% by 2020. According to statistics from the National Statistics Bureau, China had 660 cities at the end of 2002. Of these, 138 had a population of more than one million (permanent residents only), 23 had a population of more than two million and 10 had a population of more than four million. Around 280 cities have a population of between 500,000 and one million. The total land area of the cities is about 4.4 million square kilometers, accounting for 45.9% of the total area of the country. The current level of urbanisation has now reached 39.1% from the 1980 level of 19.3%. It is estimated that the level of urbanisation in China will reach 45% by 2010 and 60% by 2020. In 2003, the total population of China was 1.292 billion, of which the urban population was 523.7 million, accounting for 40.5% of China's total population. The urban population is forecast to increase to 711.7 million by 2020 and this will lead to a huge demand for transportation and energy.

Growing Automobile Industry

With China's industrialisation and urbanisation, industries such as automobiles, real estate, information technology (IT) and energy are being spurred to greater development. With the improvement in people's standard of living and consumption, the urban inhabitants' demand for automobiles is rising. The momentum in growth in the automobile industry is strong in China. Automobile manufacturing is expected to become an important pillar industry in China in the coming decades. China is now the fourth largest automobile producer and the third largest automobile consumer in the world. The output of automobiles increased to 3.25 million in 2002 from 1.06 million in 1992, growing during this period at an average annual rate of 14.8%. By the end of 2003 the production of automobiles (including cars, trucks, etc.) reached 4.4 million, up 36.7% on the previous year. Of this total, the output of cars reached 2.02 million, up by 85%. The total ownership of private cars in China reached 4.89 million, an increase of 1.46 million on the previous year, or 42%. It is estimated that by 2010 the rate of production of automobiles will reach 45 million and 10% of households will have a private car. Moreover, China is also the largest manufacturer of motorcycles in the world. The output of motorcycles is 50% of the world's total and has regularly been above 1 million since 1997, reaching 13 million in 2003. The major growth in automobile output and sales volumes indicates that the automobile manufacturing industry has become a prime mover in the growth of the Chinese economy and in making improvements in the people's living standards and consumption structure. Meanwhile, the growing affluence of people in the cities is stimulating the purchase and use of cars and is the driving force behind personal mobility and social standing. Private car purchasers in cities are the biggest automobile consumer category.

Increasing Energy Supply and Demand

The rapid growth in the use of motor vehicles in China is resulting in a huge market demand for energy. China has become the second largest energy-consuming country after the United States. In 2003, crude oil output reached 170 million tonnes, up by 1.8% on the previous year. Since China became a net oil importer in 1993, the import of crude oil has continued to rise with imports now accounting for one-third of total domestic consumption. The import of crude oil increased to 91.12 million tonnes in 2003, up from 9.37 million tonnes a decade earlier and up by 31.3% on the previous year. Refined oil product imports reached 28.24 million tonnes in 2003, up by 38.8% from 1993. It is estimated that China's demand for oil will amount to 650 million tonnes by 2020. Demand for natural gas will amount to 170 million cubic meters by that time.

Transport energy consumption will increase considerably as a result of the quadrupling of GDP and auto output over two decades. It is forecast that the proportion of transport energy consumption will increase from 11.1% in 2000 to 16-17% by 2020. In China, current known oil reserves are 20 billion tonnes, 1/5 of the total world reserves, and known natural gas reserves are 3.4 trillion cubic meters. According to the forecast in the report of the Minerals Resources and China's Economic Development, made by the Global Minerals Resources Strategy Research Centre, existing oil and natural gas reserves in China will be sufficient for consumption for 10 years, the final exploitation volume will be just enough for consumption for 30 years. By 2020, oil imports will exceed 500 million tonnes annually, natural gas imports will exceed 100 billion cubic meters and the dependence on imports will thus be 70% for oil (exceeding the level in the US today, now 58%) and 50% for natural gas. China is therefore increasingly vulnerable to the uncertainties of the world oil market with respect to both output and price as well as the political instability of oil-exporting countries.

Urban Transport Development Trend and Challenges in China

Along with rapid developments in China's economy, society and level of urbanisation, urban residents have made greater demands on an accessible, convenient, comfortable and safe public transport system. Due to the unbalanced structure that has developed in the process of social and economic development over a long period, the backward urban public transport infrastructure and existing public transport systems cannot meet the increased daily travel demands of urban inhabitants. The conflict between the supply of, and demand for, urban transport is obvious. The problem areas are as follows:

Inadequate Public Transport Systems

The weak existing public transport system in China cannot meet the needs of modern urban rapid transit. With the acceleration of urbanisation, the need for public transport is on the increase. China now has 660 cities and there are 450 cities with populations over 500,000. The permanent urban population is about 500 million. Public transport has become the first choice of transportation for the majority of Chinese city residents. However, development of urban public transport systems cannot match demand on the city level. Until now, the transport structure in most of the big cities has been isolated and lacked transport tools for coping with large numbers of passengers. Apart from certain megacities and other big cities such as Beijing, Shanghai, Guangzhou, Tianjin and Dalian, which now have metro systems in place, metro systems are under construction in Chongqing, Nanjing, Wuhan and Shenzhen. Other cities generally need to rely on conventional buses and trolley buses to carry urban passengers.

Insufficient Public Transport Capacity and Worsening Service Quality

Existing public transport capacity is obviously insufficient. Many big cities cannot meet the basic travel demands of city residents. Although some local governments have increased their financial input into urban public transport in recent years, transport capacity is still inadequate. By the end of 2002 the per capita road area in Chinese cities was 6.34 square metres; the actual number of public buses (trolley buses) in use totalled 223,572 sets, i.e. 6.79 buses per 10,000 people and there were 778,791 taxis. The total urban passenger transport volume carried by public transport systems reached 343 million passenger trips in 2002. At present, the proportion of travel by public transport in Chinese cities is less than 10% and about 26% in the megacities. However, in many cities in developed countries, Paris for example, this share has reached 65%.

The quality of existing bus services is very poor because of insufficient bus lines, worn-out vehicles and low speeds as a result of operating in mixed traffic and unreliable schedules. The average travel time, for example, is 58 minutes for Beijing residents. Because cities have reduced bus subsidies and increased fuel prices over the last few years, most urban bus companies have lost money, eliminated monthly tickets and begun to cut services. Many passengers therefore pay higher fares and receive low-quality service. Moreover, because public transport capacity cannot satisfy the demands of the large flow of passengers in urban areas, individual means of transport have expanded too rapidly and thus heightened pressure in urban traffic. This situation is particularly serious in large cities and has even become a prominent social problem, impeding the economic development of many cities.

Rapid Growth in the Number of Motor Vehicles

Amazing economic and urbanised development has brought about a rapid growth in the number of motor vehicles. If one takes Beijing as an example, it took 20 years to increase vehicle ownership from 10,000 to 100,000; it took 48 years to increase the number of motor vehicles to 1 million; but it took just six years to increase from one million to two million. By the end of January 2004 total motor vehicle ownership reached 2.15 million in Beijing, of which 950,000 were private passenger cars, accounting for 44.3% of the number of total automobiles in the city. It is expected that by 2008 Beijing will have 3 million motor vehicles.

According to the statistics gathered by the China Automobile Industry Association, in the first half of 2004, the output of cars in China was 1.24 million and sales were 1.13 million, up 36.37% and 31.59% respectively. The share of cars in total automobile production accounted for 47%, up by 2% on last year. It is forecast that automobile output will exceed 5 million in 2004 and by 2005 total care demand will reach 5.9 million, exceeding the car demand in Japan and thus becoming the second largest auto consumer in the world, reaching 8.7 million by 2010. The continued increase in private car ownership plagues urban transport systems and challenges China's sustainable urban transport development socially, economically and environmentally.

Urban Sprawl

Due to continuing urban sprawl, the average commuting distances to work and education are increasing. More and more trips are over long distances, which are beyond the range of traditional non-motorised transport modes, i.e. bicycles. Public transport has become an essential mode of transport for many urban residents. At the same time, the growing population of senior citizens is totally reliant on public transport.

Serious Traffic Congestion

Although increases in private car ownership and usage have brought city residents many advantages, they still create huge pressure on urban transport systems. Traffic congestion that is producing gridlock situations has become routine in major cities. Because of and after the SARS epidemic in particular, most large cities experienced substantial surges in private vehicle ownership although even before SARS all large cities had experienced rapid expansion in private vehicle ownership because of a rate of economic growth that has averaged 9.4% in recent years.

Meanwhile, the roadway improvements that have been made in most cities have led to a vicious circle in which building more roads has led to further traffic congestion in the city core areas because of rising vehicle ownership. In some large cities average operating speeds have been reduced to 8-10 km/hr in the central area and sometimes to less than 5 km/hr. Experts estimate that a reduction in public transport speeds by one km/hr means a loss of combined capacity of 300 vehicles. In Beijing, the 'on-time' reliability rate of public buses and trolley buses fell from 70 per cent in 1990 to 8.4 per cent in 1996; the number of buses in operation was reduced by 1,000 vehicles/day and daily bus person-trips fell to 200,000. Statistics indicate that the total economic loss is more than RMB Yuan 100 billion in urban areas. It is estimated that because of traffic congestion, the annual increase in automobile fuel consumption exceeds 300,000 tonnes and that annual automobile pollution emissions are increasing by 15% - all this resulting in approximately RMB Yuan 15 billion worth of economic damage each year.

The Weak Urban Public Transport Sector and Inadequate Governance

Planning for urban transport in China started in the 1980s and it has attracted more attention in recent years. However there is a lack of knowledge of regular urban transport patterns and inadequate techniques are employed in the making of transport plans allied with only vague conceptual ideas. Current planning for urban transport in China is therefore not sufficiently scientific and rational and it thus has a great impact on the efficient construction of urban transport infrastructures and the formulation of good transport policies. Meanwhile, the stipulation and implementation of urban transport plans is also greatly affected by political input, local economic influences, available resources and other factors. What is more, due to a lack of continuous governance, some politicians in cities are quite free to implement unwarranted authoritative transport plans and select transit options without reference to planning stability, serious thought and sustainability. This certainly affects the sustainable development of urban transport.

Negative Impact on the Urban Economy and Urban Transport Development in China

Most developed countries in the world accept the concept of sustainable development based on their experience of industrialisation. China, as a newly emerging developing country, is at the stage of speeding up its development. The huge expansion in demand for urban transport has prevented sustainable urban development from matching it and has resulted in intense pressure on China's social, economic and ecological environment systems. The issues that give rise to problems are discussed below:

Energy Security

The expansion in car ownership has increased the demand for energy in recent years. Automobile energy consumption now accounts for 6.2% of end-use energy consumption in China. Motor vehicles in China are responsible for 90% of the total petrol consumption and 45% of diesel consumption. Motor vehicles consumed 35 million tonnes of petrol in 2000. It is forecast that the average annual growth rate in petrol consumption will reach 6.35% by 2010 and will increase annually thereafter by 5.7% from 2010 to 2020. Motor vehicles consumed around 35.7 million tonnes of diesel fuel in 2000. It is estimated that the average annual growth rate for diesel fuel will reach 7.8% by 2010 and then increase annually by 7.2% from 2010 to 2020. According to a study made by Tsinghua University, the oil demand from road transport will be more than 20% of the total demand by 2020 and almost 90% by 2030, making it the principal source of oil demand and oil imports at that time. International experts predict that by 2020 China's dependence on oil imports will reach 70% of its total consumption. China is facing severe challenges in energy security and destructive air pollution caused by auto energy consumption.

Vehicular Air and Noise Pollution

The growth in private automobile ownership has not only increased urban noise pollution but also noxious emissions so that automobiles are now the chief source of urban air pollution, contributing to greenhouse gas emissions and resultant global warming. According to a United Nations report, over 80% of CO and 40% of NOx in Beijing, Shanghai and Guangzhou comes from automobile emissions. Experts estimate that automobiles in China's large cities are on average responsible for 60% of CO, 50% of NOx and 30% of HC emissions. Even though the density of cars in Chinese cities is lower than in developed countries, its motor vehicle emission discharges are similar to those in the United States during the late 1960s and early 1970s. Current pollution indices in Chinese cities are 10-20 times higher than those in Western countries.

With the increase in the number of motor vehicles, Chinese city residents have increasingly become victims of auto noise pollution. In Beijing, for traditional planning and historic reasons, one million residents or 16% of the total city population, live beside a main trunk road. The noise in Beijing maintains an average level of 71-72 decibels for a long period of time. In order to reduce noise pollution the Beijing Municipal Government has decided to invest a total of RMB Yuan 1 billion from 2002 onwards to build soundproofing facilities that are expected to reduce the noise level to 69.5 decibels by 2007.

Health Effects of Air Pollution

In the process of rapid economic development within the last decade, the vehicle population has increased dramatically. Pollution related to traffic exhaust emissions is getting worse in some big cities in China. Pollutants such as NOx, CO, VOCs and fine particle emissions, which come in large amounts from vehicles, are a major threat to human health and ecosystems. Serious air pollution causes premature death, cancer, human growth effects, asthma attacks and bronchitis. Diseases of the respiratory system caused by air pollution are the third largest cause of death in Chinese cities, claiming 330,000 lives each year. According to a recent study, the medical cost for cancer patients is about RMB Yuan 80 billion in China and accounts for 20% of the nation's total medical costs.

Traffic Safety Concerns

Urban traffic accidents have caused a large number of casualties and huge direct and indirect economic losses. Official statistics indicated that there were 74,000 road traffic deaths in 1997. This means that the accident rate for motor vehicles is 46 per 10,000. This is almost twice as high as that of India, six times that of Thailand, 23 times that of Japan and 40 times that of the Northern European countries. In 2002, China suffered 773,137 traffic accidents, causing 109,381 deaths and 526,074 injuries, resulting in RMB Yuan 3.32 billion worth of direct economic loss, corresponding to 4.24% of the GDP. This means that every five minutes someone was killed in a road accident in China in 2002. Today's figures are very likely much higher.

Toward Sustainable Urban Transport in China - Bus Rapid Transit (BRT) Systems

Transport development experience in many cities worldwide show that it is impossible to solve urban transport problems by building more and more urban roads although many newly built roads might help meet the demands of vehicle growth in the short term and ease road transport pressure temporarily. However the increase in road space can never match vehicle growth rates. Some city governments have not accumulated sufficient experience and capacity to deal adequately with the complexity of urban transport development. Some large cities have planned to build metro systems (other than those they have now) to facilitate public transport. However, the high initial investment costs and consequent high operating costs have led to slow or delayed implementation. In the face of urban transport problems that need to be solved urgently, most Chinese city authorities are realising that the car is not the solution – it is the problem. They are beginning to seek new sustainable solutions to promote public transport systems and they are seeking a new and affordable public transport mode to replace and supplement metro systems. Bus rapid transit (BRT) is a great opportunity for most cities.

Advantages of the Bus Rapid Transit System

The major advantages of implementing BRT are lower initial capital investments and annual operating costs as well as offering a short construction time. BRT systems can offer capacity that is closed to a metro system. It is estimated that the construction cost for a BRT corridor is between RMB Yuan 20 million and 100 million per kilometer while it would cost RMB Yuan 400 to 700 million to construct one mile of metro. In general, it would take less than 18 months to construct a BRT corridor. BRT is a very flexible mode of public transport which can not only link up to multi-modes of public transport but it can also become a transit or supplement to metro and light rail. Meanwhile, it can introduce modern ITS techniques and facilitate energy savings and environmental protection. What is more, BRT systems can provide high-quality, customer-oriented transit systems delivering fast, comfortable and low-cost urban mobility. The experiences in Curitiba and Bogotá in South America show us that it is possible to satisfy the needs of middle-income and low-income groups and entice passengers from their private cars. BRT is therefore a cost-effective, equitable and environmentally friendly transport mode. It is a very attractive transit option for a developing country like China. The extensive application of BRT in Chinese cities will promote sustainable urban transport development socially, economically and environmentally.

Initiatives

Despite the significant advantages of BRT systems, BRT is a new concept for most decision-makers and city planners in China. The Institute of Comprehensive Transportation (ICT) has been spear-heading BRT promotion efforts since 2000. In association with international agencies, ITC has carried out many BRT development initiatives, including the following:

With financial support from the W. Alton Jones Foundation, ITC conducted the "Green Transport for Cities in China" symposium. The ICT and the Urban Transport Centre (UTC) of the Ministry of Construction also conducted various BRT workshops/seminars in Beijing, Xiamen, and Taipei. Technical meetings have also been organised, in response to local needs, to plan BRT and transportation management in selected cities such as Wuhan, Shanghai, Xian, Kunming, Chengdu, Chongqing and Xiamen.

The "Green Transport Symposium for Cities in China" was held, with strong support from UTC and ICT, in 2001. Over 60 participants from 20 cities attended. Delegates representing transportation planning, traffic enforcement and management, urban planning as well as environmental protection agencies attended the sessions. The number of cities participating and the number of participants from the various cities far exceeded the original expectations. These cities included Beijing, Ningbo, Changchun, Luoyang, Nanjing, Harbin, Anshan, Zhuzhou, Guiyang, Shijiazhuang, Changsha, Wuhan, Nanjing, Qingdao, Xiamen, Shenzhen, Shenyang, Zhongshan, Xi'an, and Zhengzhou. Of these, Ningbo and Changchun are two cities that were involved in the previous phase of this project. Many of the delegates are decision-makers in charge of their respective government agencies.

Based on recommendations from the ICT, a BRT brochure was produced with the help of an international non-governmental organisation. The brochure discusses the BRT concept and the potential roles and development strategies of BRT systems in a city's public transportation system. The brochure provides useful material for educating decision-makers about the benefits of implementing BRT systems and provides technical guidelines for planning and implementing BRT systems in Chinese cities.

ICT also co-organised the Beijing BRT Development Symposium in March 2003. Jaime Lerner, the founding father of the BRT concept and former Mayor of Curitiba, Brazil, participated in the symposium. This symposium was an inspiring start to BRT development efforts in Beijing. Under the auspices of the ICT, the head of the National Development and Reform Commission (NDRC) of China, led a delegation of mayors on a study tour to Curitiba in Brazil, and Ottawa and Vancouver in Canada in November 2003 to learn about the best BRT practices. Mayors who have become well aware of the advantages of BRT have expressed a strong desire to build BRT systems in their own cities. Meanwhile, the head of the NDRC has shown great interest in BRT and has presented modern experience of BRT from abroad to mayors in China, including the mayor of Beijing. He has suggested introducing the "priority to develop BRT in Chinese cities" into "The 11th Five Year Plan".

BRT Practice in Chinese Cities

After a series of education initiatives related to BRT and the concept of sustainable urban transport delivered to city decision-makers and technical staff, many city governments are beginning to accept the BRT concept. Kunming is a good example of BRT practice.

Kunming, a world-famous tourist city, opened the first Bus Rapid Transit system in China during the International Horticulture Fair in 1999. The BRT system is 5 km long and is located on the most important north-south trunk line (Beijing Road) within the major trunk road network in the city, crossing the core areas of the city. A large number of office buildings and commercial and financial centres are concentrated along both sides of the BRT system. It has become a core hub axis and a main passenger transport corridor and has greatly improved the service standard of public transport in Kunming. Since the system was brought into operation in 1999, the daily bus passenger traffic volume increased in 2001 and 2002 to 750,000 trips/day from 500,000 trips/day, the average operating speed has risen to 15 km/hr from 9.6 km/hr, and the satisfaction rate of passengers is over 96%.

The good pilot result and role of BRT in Kunming is making the municipal decision-makers realise the advantages of BRT as an effective public transport option in dealing with urban traffic problems. Meanwhile, the success of the pilot system has strengthened confidence and determination on the part of city politicians in giving priority to developing public transport. Based on a light rail network in city transport planning, Kunming will build a BRT network of 42 km with the city centre as its core. After the BRT network is built, the residents will be no more than 300 metres from any BRT station in 75% of city centre areas. This will provide greater convenience for passengers. The proportion of bus passengers would be expected to reach 20%-25%. The service levels of public transport will play an important role in constraining private car transportation and improving travel structures and travel behaviour. A strong public transport system will be helpful in building up a sustainable, environmentally friendly transport system with a high level of efficiency and low energy consumption.

The experience in Kunming proves that merely providing a complete public infrastructure is not enough. "Public transport priority" should be an integrated development strategy. In order to realise the goals of accessibility and high efficiency, certain corresponding measures will need to be implemented, such as disseminating a policy for "public transport priorities" to the public and gaining their common understanding and support for a number of necessary changes. Such changes would encompass limiting private car traffic in central areas, providing parking places and introducing parking fees, optimising road network structures, including bicycle lanes, giving priority to public traffic control signal changes, applying GPS in bus control, and introducing a ticket fare reform.

Kunming has set up a very good example of BRT development. Some megacities, such as Beijing and Shanghai, and large cities such as Chongqing, Chengdu, Shijiazhuang, Xian and others, have already begun to take action to develop BRT in their cities with preparatory work already under way.

Beijing, as a megacity and host to the Olympic Games in 2008, is facing many pressures and challenges. Stress from urban traffic congestion problems is caused by increases in travel demand and growth in private car transport that is too rapid. Related to these is an energy shortage and environmental pollution. At present, the public transport system in Beijing is relatively backward. Previous attempts and efforts to improve bus transit have been constrained by serious traffic congestion. The metro has not been able to accommodate increasing travel demand. In order to meet the traffic requirements of the Olympic Games in 2008 and to be better able to deal with the complexity of urban transport development, the Beijing Municipal Government has been seeking a sustainable, cost-effective and environmentally friendly public transport option. Through a series of BRT encouragement activities, the mayors in Beijing have recognised the advantages and roles of BRT. They clearly realise that it is impossible to deal with the present traffic stresses merely by improving regular bus transport and facilitating more metro building.

In 2003, the Beijing Municipal Government made a decision to build the first BRT pilot corridor in the southern middle axis road of Beijing. It runs from Qianmen via the Temple of Heaven to Demaozhuang, with a total length of 15.8 km. In order to implement the BRT corridor successfully, the Beijing Municipal Government adopted a set of effective measures. These include adjusting the existing transport development plan to incorporate BRT, adopting the strategy of "outside to inside" in BRT corridor design planning, paying more attention to the integration of BRT corridors with other modes of public transport and stressing a combination of land use management and BRT development. The feasibility study and planning phases have now been completed and construction is expected to begin early in 2005. The completion of a BRT corridor in Beijing will set a good example in integrating multimodal public transport, including metro and regular public transport. This pilot project will have provided a model for other cities, especially for the poor western cities that cannot afford to build metro or light rail but which urgently need to improve their urban public transport systems now.

BRT is currently a new mode of public transport in China. It is being widely adapted for use in many Chinese cities. In order to promote BRT development in Chinese cities, it is necessary to develop a National BRT Development Policy and establish National BRT Technical Guidelines and Norms. In the future, the ICT will strengthen its co-operation with international agencies and foundations pertaining to BRT development in both technology and investment needs.

Conclusion

Worldwide problems of urban sprawl, urban traffic congestion, energy shortages and environmental deterioration are plaguing most Chinese cities. Along with the rapid development of urbanisation and motor vehicle use, China, as a developing country that has just lifted itself out of poverty and backwardness, is looking for a better public transit system to attract more passengers from private transport modes and to meet increasing transport needs. Metro and light rail systems are desirable but the majority of Chinese cities, especially the underdeveloped cities in the western regions, cannot afford these transit options. Bus Rapid Transit (BRT) is a cost-effective alternative that has been developed in many developing countries. Successful experience in their use in some Chinese cities has proved that BRT is a sustainable urban transport mode that will contribute to sustainable urban transport in Chinese society, the economy and the environment.

References:

1. Year Book of China Transportation and Communications Statistics in 2002 and 2003
2. China Development Report in 2002 and 2003
3. Modern Exclusive Busways – Practice, Review and Outlook of Kunming, a Lin Wei Paper in Symposium on Beijing BRT Development Strategy, 2003

Chapter 8. The Complex Relationships between Motorized Mobility, Societal Development and Sustainability

Marie Thynell

Introduction

The impact of mass motorisation on societal development is a neglected area of research within my field, International Political Economy and Development Studies. In particular, the social component of development in the upgrading of transport systems is largely omitted. I will therefore focus firstly on the emergence of motorised transport systems in two very different cities, secondly on the handling of current urban transport problems in two other very different cities and thirdly on the role of the search by certain global corporations for strategies to confront the everyday problems of urban traffic. Finally, some general comments are made about what kind of transport policies are seen to contribute to societal development.

My theoretical framework is based on a combination of theories in International Political Economy and the theory of Large Technical Systems. The methods used in the empirical inquiry are interpretation of statistical data, structured interviews (with key politicians, NGOs and transport administrators), text analysis and participant observations. The results have been published in my recent thesis *The Unmanageable Modernity. An Explorative Study of Motorized Mobility in Development* (based on a study supported by the Volvo Research Foundation).

The problems of motorised transportation are well known. My research topics are defined against a background of six categories of negative consequences of current mass motorisation:

- Injuries (1,200,000 people are killed and an estimated 23-34 million are injured on the roads every year). A large number of other health effects should also be added to these figures
- Negative effects on local and regional ecosystems and on the global climate
- Negative effects on the man-made urban environment: cultural and historical legacies are damaged and public areas in which to socialise are reduced
- Increasing congestion
- Possible future difficulties and potential conflicts in providing the rapidly expanding global fleet of motor vehicles with adequate fuel (the forecast number of vehicles in 2030 is 1,630 billion.)
- Finally, (and I want to stress this effect) the social division between those with access to motorised mobility and those without. In societies with huge groups of poor people this effect impedes considerably the potential for economic growth and improving the life quality of individuals. Considering the fact that – in a global perspective – the number of poor people is growing and the fact that the gap between the haves and the have-nots has increased, these negative consequences are very serious from the point of view of societal development. In 2003, two billion people are estimated to be poor and according to the UNDP approximately 70 per cent of these are women. It is therefore of relevance to pinpoint the way in which mass motorisation relates to societal development beyond the spectacular economic investments in infrastructure that are being made

A historical survey

My historical study of the emergence of motorised mobility in two very different social and cultural contexts, Brasilia in Brazil and Tehran in Iran, stresses the strong hegemonic influence of Western countries regarding the development of motorisation in general and automobility in particular. The motorised transport systems in the U. S. influenced the development of the car culture in both Brasilia and Tehran. By this I mean the vision that individual mobility should be private and motorised. In the original urban planning of Brasilia, from the 1950s, car use was given priority in an extreme way. In Tehran, an important part of the transport system was in the process of being automobilised during the 1960s and 1970s as well as during the time of the Islamic Republic. In Tehran today huge investments are being made to upgrade the car and road system and to increase the number of motorised vehicles. Huge investments are also being made in public transport (underground) but far from sufficient to provide reliable access for the approximately 16 million inhabitants of Tehran. Consequently, the most important investments in the two cities have facilitated automobilisation and the individual mobility of more affluent inhabitants.

A comparison between Rome and New Delhi

My study also shows the limited role of local interpretation of the problems and the limited local influence on motorised mobility as a part of the modernisation process. (Innovative ways of coping with accessibility – adapted to the needs of developing countries – have rarely been spread globally, if ever). Evidently, motorised mobility is a core activity in today's cities and mobility is intimately related to the dynamics of economic growth according to the global and universal dimensions of motorisation. My study of how current transport problems are perceived and handled in two very different cities (Rome and New Delhi) has identified significant differences among the political and professional actors as well as among representatives of NGOs in the two cities. A number of car-related phenomena – such as problems of noise, safety and related health problems – were commented on. The problems are, however, expressed differently in the very rich and historical city of Rome and in the poor but rapidly expanding mega-city of New Delhi. This can be seen with regard to health problems for example (bronchitis in New Delhi and sedentary-related heart problems in Rome). In Rome, the loss of space for socialising was often commented on together with the decay of historical monuments. This was less important in New Delhi, where accessibility and the lack of an efficient public transport infrastructure were the major topics. Meanwhile, the problems of bad public service and non-existent timetables (sic!) constitute a problem that the two cities have in common. There also seems to be a common and firm belief in the ability of technological development to solve the current problems of transport systems in both cities.

The role of the automotive and oil industries

Bearing this in mind, I decided to turn to the automotive and oil industries to see how they perceive the problems of urban transport and what they consider to be appropriate strategies to improve urban traffic conditions. I made an analysis of written policy documents from three leading automotive companies and one large oil company. The results show clearly that these actors (with some minor exceptions) only focus on the development of their products and on the adaptation of their own production, not on the necessary changes in the entire motorised system in cities. It would therefore seem hazardous to rely only on technical development as the solution to the complex problems of urban transport development.

Some conclusions

An important but worrying conclusion is that there are few signs of reflexive thinking among representatives of the state or of the market about how to handle the fundamental crisis of motorised transport systems and their negative side effects. The continued expansion of motorisation has brought urban transport to the verge of a crisis that none of the traditional actors seems to be able to manage. So far, none of these actors have addressed the whole system as such. Instead, priority has been given to issues of traffic efficiency, to lowering the cost of transportation, and to various technical solutions regarding traffic flows, driver safety, emission control and cleaner production processes. It is therefore fair to claim that the private and public builders of motorised systems have largely neglected the necessary system changes despite the impressive negative side-effects these systems have on the environment as well as on social equity.

It can also be said that politicians have largely failed to regulate the market expansion of mass motorisation by means of political inventions. This trend is nowadays aggravated by the fact that 50 per cent of the global population live in cities and that about 10 per cent of the entire global population own a car. The vested interests of the various actors are reflected in the socially constructed policies and expressed in the politics of mobility perspectives. There are different and conflicting perspectives regarding future urban mobility and in some cities the transport policies have set in motion a shift from the use of the traditional toolbox of policy instruments towards an integrated view of urban mobility, such as in New Delhi in 2003.

There is therefore room for many new initiatives to create a knowledge base for urban transport systems that is efficient, safe, sustainable and compatible with human, social and economic development. I will highlight some aspects of such studies, which seem to be important in this perspective. One task is to increase the number of case studies of cities in both developing and developed countries and to deepen the comparative analysis of these cases. Future research should address the strategies of the politicians to increase sustainability and embrace ways to overcome the various barriers to the implementation of these strategies. This relates both to the development of technological systems and to the current phase of market globalisation.

Studies of policies for the future

The main focus in the future should, however, be on the politically guided process of a system change. Who are involved in managing motorisation, internationally, nationally and locally (particularly in mega-cities)? What do they define as their target locally? What, for instance, is their view of environmental and social sustainability? In addition, there is the issue of who defines the problem and elaborates on the strategy. Is, for instance, the problem in question a technical, economic or social issue also expressed as the social construction of mobility perspectives? How are issues of poverty, age, health and social equity dealt with? What links to other societal sectors do local and international politicians pay attention to? The definition of problems and strategies differs on the various levels and but which will be the winning interpretation? From the beginning the issue of emissions seemed to be the problem of wealthier countries. What about the role of global automotive and oil corporations? What are the strategies (content and direction) of national and local political actors? One of my research experiences is that – on the city level – there are often various barriers that impede the implementation of decisions taken on the national level, due to a local need for economic and political compromises and considerations. How is the complexity of the changes handled in different developing cities, when the market and the state dominate the decision-making process and civil society is being marginalised?

In the analysis of the local processes of change there are variations in background and dynamics that make it relevant to specify the local features influencing the processes of change. These could be geographical, political, economic, social, cultural and other special circumstances of a local character. Global cities are inserted differently into globalisation and are characterised by a huge capital flow in certain sectors, highly educated people, the latest innovations in information and communications technology, the presence of multinational business, growing formal and informal economies and the considerable increase in the use of cars. Automobility is seen as a necessary condition for individual and rapid access to the various parts of the cities but also as a core part of the modern lifestyle. In order to come to grips with the local character of motorisation and transport needs it is fundamental to understand underlying structural relations that contribute to the shaping of modern mobility.

There is interplay between local and international transport relationships. The political pressure emerging from climate change might be an international aspect that has an influence in some cases and not in others depending on local conditions, national law, the profile of the vehicle fleet and so forth. The practice of fuel taxation can be connected to cars as vehicles of economic growth as shown by the World Bank in a Central Asia transport project. In the Central Asian republics for instance, the definition of the problem comes from the outside or, expressed differently, as a top-down transference in which the strategy has emerged in a strikingly different socio-economic context. The external influence on development in the motorisation sector and urban development co-acts with national conditions and ways of coping with this development. They are also closely connected to huge economic investments. The taxation of fuel is often based on local traditions concerning taxes, views on energy and so on, whereas the early CO_2 debate often reflected global environmental politics rather than locally perceived difficulties.

The theoretical basis of continued studies

In order to analyse processes of this kind I am revising the theory of Large Technological Systems and linking it closer to the theories of International Political Economy, i.e. to analyse dominant actors from the perspective of economic and political structures. (By this I refer to the underlying political order and market relations.) So far, theorists of Large Technological Systems have mostly dealt with the expanding systems and with system problems such as 'reverse salients' in order to overcome the challenges. They have not handled situations where technical systems approach a fundamental crisis in their historical development. The analysts have also been seriously criticised for identifying themselves too much with the 'system builders' and with the system itself. Hence, due attention has not been paid to the socio-economic groups excluded by the technological system in question, such as the implications of economic exclusion of social groups. This exclusion creates divides within cities and from the point of view of societal development exclusiveness creates imbalances and fragmentation. Despite their marginalisation, excluded groups are part of societal development and also interfere with technological systems. This is proving to be a considerable problem in some cities since 50 per cent of the passengers do not pay the fares (according to the World Bank), the majority of the trips are paid by social welfare or the passengers are employees subsidised by their employers.

In welfare states or modern nation states motorised mobility and the relationship between the state and the market is determined in part by citizens as both voters and consumers. The lifestyle of the consumers is therefore an important economic factor that sustains the market economy. In the future, conflicts about lifestyle, car use and consumption will probably be intensified.

Consequently, the role of society in the process of change needs to be included in the study. There is still a knowledge gap to be bridged regarding the way in which transport policies and economic investments contribute to societal development in different parts of cities. This could be changed, for instance, by evaluating the performance of technological innovations in terms of social behaviour, economic changes and the technical outcome (e.g. a shift to CNG, road pricing, bus licensing). Many projects have not been evaluated in an appropriate way. One reflection concerning transport changes is that isolated projects might not be sufficient – the dominant actors, such as the World Business Council for Sustainable Development and the World Bank, seem to benefit from implementing integrated programmes, including poverty reduction (PRSP) and the Millennium Development Goals (MDGS), rather than isolated efforts.

In developing countries investments in expensive transport systems often lead to splintered access. Sometimes the same investments lead to less societal development. A closer look at the dynamics of accessibility of vulnerable groups in a bottom-up perspective is often necessary to confront social exclusion and increase the positive trade-offs between investments and development. In the ideal case huge infrastructure projects (metro, BRTs, etc) are evaluated from the point of view of improved access for vulnerable groups. Integrated approaches, such as the ones put forward by the World Bank (PRSP; MDGS) and others, are helpful in overcoming both lack of transport and poverty.

Today's problem of urban transport has both a structural base and an actor-orientated dimension. To what extent is the urban government in control of the transport system? In expanding mega-cities the decision-making process regarding land use and the location of industries sometimes has a weak or inefficient political institutional background. In poor countries the public sector cannot maintain or invest in the infrastructure and national finances do not usually support large infrastructure improvements.

In all studies on how to cope with the complexity of urban transport development one could bear in mind that the car has become one of the leading objects of our culture. The car is today also an economic harbinger, probably most visible as a sign of wealth in developing urban areas. The French theorist Michel Callon expresses this view in the following words:

> The automobile is at the nerve centre of society, so socially embedded that it can be modified only with great care (Callon 1993:89).

This phrase stresses the fact that the interests of the market actors and political actors often converge. This calls for a politically guided strategy to confront today's increasing problem of urban mobility. What are the political strategies to confront the need for economic growth and increased physical mobility? Hopefully such strategies will be linked to an inclusive notion of managing mobility and a comprehensive vision of future urban transport that is linked to genuine societal modernisation.

Section three: Coping with Complexity: Some Analytical Perspectives

Introduction

The third section of this anthology includes a number of chapters, in which the authors analyse the difficulties of handling the complexity of urban transport development and in which they present some concrete examples of how urban transport systems (or parts of these systems) have nevertheless been managed in a more or less constructive way.

In the first chapter, Robert Cervero describes some powerful megatrends that today influence the mobility marketplace in the US. He points to the fact that circumstances such as shifting socio-demographics, shifting economics together with changing technologies and shifting urbanization patterns reinforce each other, and that they are "conspiring against all forms of movement except the private car".

Then Cervero draws attention to the complexity represented by the experts' conflicting ways of interpreting the problems of urban transport. This complexity has to do with different conceptual frameworks, different operational assumptions and different analytical styles. He illustrates this observation by describing a part of the debate on induced travel demand.

As a third step Cervero refers to the complexity in the political and institutional landscape. News ideas in the field of transport and mobility are often "obstructed by messy institutional landscapes and political detours". Coherent visions of future land-use and transport systems are rare, and the role of visionarics is unusual in the US (there are however some good examples outside of North America).

In the final part of his chapter, Cervero describes some promising initiatives in the US. He refers to increasing interest in car-sharing and in shuttle programmes for trips to and from the workplace. Such initiatives help to reduce the need for private cars (and for a second car in the household). Programmes for public transport-orientated development contribute to reducing traffic congestion as well as to reducing urban sprawl in some places (Portland, Oregon is one of these places – see below Chapter 14). Now institutional arrangements are also of some importance. In the state of Georgia, for instance, a Regional Transportation Authority has been formed.

Finally, Cervero concludes that there are no reasons to predict any great change in the transport systems of the US in the near future. "Middle-class and well-to-do households with several children and a preference for privacy and seclusion will continue to reside mostly in the suburbs and beyond". The reader may, however, make the observation that a question never raised in the chapter is, what kind of consequences a future major and lasting oil-crisis will inflict on the US transport system.

In the following chapter, Geetam Tiwari discusses some of the issues dealt with by Cervero (present mega-trends and the experts' ways of interpreting the problems) but in a very different socio-economic and cultural environment. Her starting-point is that the Asian region has a potential for rapid urbanization in the near future. This growth has very much to do with an ongoing unplanned immigration of poor people to the cities. These poor sections of the urban populations remain, as a rule, outside formal planning processes. Their economic activities (aiming at subsistence) are mostly informal. At the same time, the growth of the formal economy is partly associated with the growth of the informal economy.

Geetam Tiwari then describes the traffic situation in many Asian cities. In order to carry out their low-income jobs within the formal economy, or to be able do manage their informal economic activities (for instance as street vendors), poor people need to move. This necessary mobility is carried out through walking, cycling and, sometimes, by para-transit (e.g. cycle rickshaws) or by over-crowded public transport. Indigenously designed and produced vehicles (including motorised Three Wheeler Taxis) dominate in cities lacking adequate public transport. This mode of transport requires minimal training to operate. The use of these vehicles may, however, violate formal plans and neglect recently introduced safety standards.

At the same time, the fleet of motorized Two Wheelers is growing in many Asian cities. Car expansion has also been initiated in quite a few cities, where the Asian middle class is growing in number and wealth. One should not forget, Tiwari claims, that in many Asian cities, the streets "are not restricted to movement of vehicles and people only. The range of activities includes services required by the users and social activities which require safe public spaces."

Asian formal transport policy and planning very often disregards these facts, and if they are attempted, bans on certain kinds of vehicles are issued. These bans often have no meaning for people without choices but striving to survive. In the future, Tiwari claims, one should therefore integrate not only the transport situation of the poor in transport planning but also their ability to establish self- organising systems mostly built on zero-pollution vehicles – bicycles and rickshaws. What they need is infrastructure for non-motorized vehicles. Such a transport policy and transport planning would "meet the contrasting needs of different social groups". The poor social groups cannot be wished away. "They are here to stay".

In the next chapter, Vinand Nantulya, Meleckidzedeck Khayesi and Wilson Odero discuss some examples of positive changes in complex urban transport problems in sub-Saharan Africa. These positive changes are based on what the authors regard as emerging opportunities in the political arena, in the growing informal transport sector and in networking between various actors. There are also new opportunities for learning from other countries in the southern hemisphere. The challenge today is to maximize existing opportunities, engage stakeholders and secure networking.

Any urban transport system is regarded by these authors as 'complex' with respect both to the development of the entire system and to the organisation of different modes or subsystems. The different modes have different dynamics driving them. The transport system is also seen as embedded in a set of social and economic relationships as well as a part of the physical environment. All this adds to the complexity.

To understand the complexity of current problems in sub-Saharan Africa and the scope of the challenge of coping with the problems, some characteristics of the urban transport systems in Africa should be kept in mind. The transport infrastructure is inadequate and there is a mix of transport modes, and these are poorly integrated and synchronized. There are increasingly dysfunctional public transport systems and, in many cities, a growing and largely unregulated informal transport sector characterized by a multiplicity of service providers and an unorthodox competition for space and market. Finally, there are weakly articulated urban transport policies as well as poor planning and regulation. All this results in a negative image that "makes it difficult for a number of people and institutions to realise that there are unfolding opportunities in the midst of what looks like a hopeless situation". There are, however, also some good examples of political and economic reforms on the political arena. The one-party system is gradually giving way to multi-party politics. In urban governance there is a new generation of mayors and an ongoing discussion of necessary reforms. Political leaders are sometimes getting involved directly in transport issues, e.g. in Kenya and Uganda. New urban transport strategies are emerging e.g. in South Africa and Ghana. In many cases, however, planning is focused too much on motor vehicles and does not take into account the dynamics of the informal sector. What is more, it is often forgotten that Africa is – to a very large extent - a walking continent.

Finally the authors of the chapter on African urban transport mention two initiatives in other parts of the South that could be used as models, one in Latin America and one in Asia. The first model is offered by Bogotá, where initiatives taken by its mayor from 1998 to

2001, Enrique Penalosa, have reoriented the transport policy of the city in order to improve the transport possibilities of the poorest. The second model is offered by research projects and programmes undertaken at the Indian Institute of Technology in New Delhi, partly described by Geetam Tiwari in Chapter 10.

In the final chapter of this section, John Whitelegg brings the reader to Europe. His discussion is entirely practice-orientated. Emphasizing "the enormous gulf between best examples and the worst" he presents a number of best practice in cities in Germany (e.g Aachen), Netherlands (e.g. Groningen), Switzerland (e.g. Zürich), Italy (e.g. Bologna), UK (e.g London, Edinburgh, Leister), France (e.g. Paris), Denmark (e.g. Copenhagen), Spain (e.g. Madrid) and Austria (e.g. Vienna).

He finds, however, that in many European cities walking is still "of poor quality because of the priority given to vehicles". Travelling by bicycle is safe and enjoyable in only a few countries (the Netherlands and Denmark). Going by metro is more seldom a pleasure but good examples are found in Madrid and Vienna.

As a second step in his mainly optimistic approach to the complexity of urban transport problems, Whitelegg enumerates measures that have proved to be efficient. Urban planning is able to reduce the demand for motorized transport. Limitations on car park numbers can lead to a substantially reduced number of trips into city centres. Various forms of traffic management have resulted in reduced car mobility (examples are taken from Germany, the Netherlands, Sweden and the UK). Big companies and public institutions have introduced green commuter strategies in order to inspire their employees to leave their cars at home and to go by public transport.

Whitelegg also indicates a number of efficient policy instruments for achieving traffic reduction. The main instrument is fiscal intervention. He presents a long list of taxcs, such as fuel taxation, taxation on parking spaces and urban road pricing. He also recommends eliminating company car and business mileage tax regimes. Traffic reduction for lorries can be achieved through partnerships between all those involved in delivering/receiving goods in city centres.

Finally, Whitelegg draws attention to the gap between possibilities (in terms of best practices) and realities (in terms of traffic jams etc) in many cities. Recommended measures and policy instruments are often seen as efficient but not usable due to the political situation and to “the deeply ingrained negative attitudes of many citizens”.

Chapter 9. Progress in Coping with Complex Urban Transport Problems in the United States

Robert Cervero

Introduction

Complexity rears its head in many forms in America's urban transport sector. One unmistakeable outcome has been delays in advancing change and innovation. This is principally because complexity makes problem definition and resource-allocation choices difficult.

This paper explores multiple dimensions of complexity in a U.S. transportation-policy context, discusses the implications of these dimensions for policy change, and to the degree appropriate, suggests strategies that might be pursued to overcome, or at least better "manage", complexity. Three major spheres of complexity that are addressed relate to mobility markets, problem definition and analysis (technocratic complexity), and decision-making. The paper closes with a review of promising developments in coping with the panoply of complex problems faced in America's urban transport sector, with a particular focus on progress made in better integrating public transport and urbanism in the world's most car-dependent cities.

Complexity in America's Mobility Marketplace

Complexity in America's mobility marketplace is a product of powerful megatrends, including shifting socio-demographics, shifting land-use patterns, and shifting economic relationships. Of course, such "shifts" have not been independent of each other, and indeed, it is only by appreciating the endogeneity of relationships and taking a holistic perspective that one can hope to get a handle on the complexity issue.

Shifting Socio-Demographics

America has always been a melting pot of different cultures and heritages, however historically this has been represented predominantly by immigrants from western Europe, the Mediterranean region, and Africa. America's pluralism today reaches all corners of the globe, with each culture bringing with it different lifestyles, familial compositions, and consumer preferences. Cultural pluralism, while it unquestionably blurs our understanding of mobility markets, becomes an opportunity for change when one considers, for example, public transport policy. Many recent immigrants from Latin America, the Caribbean, eastern Africa, southeast Asia, and the Indian subcontinent bring with them a culture predisposition and heritage that is more accepting of small-scale, demand-responsive forms of paratransit, like jitneys, private vans, and micro-buses, even if it means sometimes tattered seats, crowded seating conditions, and not the newest of vehicles. Fixed-route, fixed-schedule buses that run on 30-minute headways are not necessarily competitive modal options to the many-to-many, on-call, flexible service attributes of the private car in the minds of many recent immigrants (not to mention middle-class Anglo-Americans). Similarly, those coming from areas with a rich tradition and spectrum of transit and paratransit offerings might be more receptive toward transit-oriented development (TOD) – i.e., living in compact, mixed-use neighbourhoods within convenient reach of public transport. From a policy perspective, the challenge is significantly one of breaking the monopolistic stranglehold that many U.S. public transit agencies and taxi franchises currently have over the mass transportation marketplace through deregulation and open market competition.[7] TOD could be leveraged through changes in traditional zoning standards and property-development codes in keeping with shifting demographic compositions of neighbourhoods.

Other examples of socio-demographic complexity and their mobility implications have been well-documented, such as increases in non-traditional households (e.g., singles, childless couples, non-related

[7] R. Cervero, *Paratransit in America: Redefining Mass Transportation.* Connecticut: Praeger Press, 1997.

adults), maturing of the population (e.g., the greying of baby-boomers), and steady feminization of the workforce. (Some projections are staggering – for example, by 2025, there will be 27 states with 20 percent of their populations over 65 or more, higher than Florida today). Smaller households with more independent members complicate the ability to form carpools and ride-matches. Older Americans might seem like natural candidates for public transport services, however the spread-out nature of many U.S. cityscapes compels many to drive. For widows and widowers, automobility is essential to avoid the social isolation all too often encountered in a spread-out, single-use landscape.

Shifting Economics/Changing Technologies

Intense competition within the global marketplace is today giving rise to modes of economic production with fundamentally different space-time arrangements than yesteryear. Post-Fordist trends toward contingent labour, sub-contracting, and flex-spec/cottage-scale forms of production have given rise to dispersed temporal patterns with profound mobility implications – e.g., unpredictable work schedules, 10-hour compressed work weeks, and flexible hours. Growth in information technologies and brokerage industries has similarly expanded the temporal envelope of commuting. Telecommunities, designed and marketed to software programmers and other "information processors", are radically changing traditional time-budget theory as we know it. For example, the city of Montgomery far north of Toronto has been designed (eg, laced with fibre optic cable) and marketed as a mixed-use community suited to telecommuters who only need to make the 100-km long trek to their main office in central Toronto once or twice a week. Facing the prospect of commuting to work as little as two to three days a week, telecommuters are willing to trade off occasionally ultra-long commutes for the quality-of-life benefits of semi-rural living. Time budgets are increasingly weigh in weekly, not daily, terms. Between 1990 and 2000, the fastest growing counties in California were not in the big urban centres but rather the foothills of the Sierra Nevada, within several hour commutes of the San Francisco Bay Area or a two-hour plane ride to Los Angeles. While living outside the boundaries of metropolitan planning organizations (MPOs), this growing tide of rural telecommuters has effectively expanded the commutersheds of big urban centres into once-rural domains. There has been an institutional lag in acknowledging this, meaning that today's geographic boundaries for strategic long-range planning are anachronisms, throw-backs to the era where people commuted five to ten miles each day to a major urban work centres. The entire infrastructure of long-range transportation planning is compromised to the extent that future "O's and D's" (origins and destinations) lie outside the jurisdiction of decision-makers who are responsible for programming long-range capital improvements to highway and transit systems.

Tomorrow's laboursheds can be expected to expand outward even more to the degree that the information-technology revolution continues unabated. Long-term impacts turn on the question of whether cyberspace, while expanding the reach of economic production, effectively shrinks physical mobility. More basically, the debate continues over whether telematics, e-commerce, and the Internet will significantly, over time, substitute for or stimulate physical travel. What is unassailable is that future travel will take on new shapes and forms: international trips (air travel) will increasingly substitute for intrametropolitan trips (car travel); with e-commerce, truck delivery trips will replace personal shopping trips; and real-time information on how to avoid congestion will enhance automobility. E-commerce could prompt the emergence of goods distribution centres in different pockets of the city. Cyber-work will likely exert growing pressures for in-neighbourhood shops, services, and "watering holds" for those wanting a break from staring at a computer screen for hours on end. Global-sourcing promises that airports and all the ancillary activities around them will become dominant activity centres and trip generators, what John Kasarda of the University of North Carolina calls "aerotropolises", the latest in the historical wave of transport and locational relationships.

Shifting Urbanization Patterns

Another megatrend closely related to shifting economics that adds another layer of complexity to the contemporary mobility market is the atomization of land-use patterns. Thinly spread, segregated land-uses – enabled by rising affluence, telecommunication advances, and a host of other de-concentrating trends -- compel people to drive, particularly in a country like the United States where by global standards petrol, parking, and (broadly-speaking) car ownership and usage are substantially under-priced. Robert Lang's recent book on the "edgeless city" contends that the drive for corporate autonomy, the location-liberating effects of cyberspace and telematics, and rising affluence in general have conspired to create a new geomorphology for economic production – sprawling corporate enclaves, business parks, power centres, and other "non-nodal" forms of development.[8] Today, all U.S. metropolitan areas (with the exception of New York and Chicago) have the majority of office space outside of traditional downtowns. While 38 percent of all office space in U.S. metro areas was located in primary downtowns in 1999, nearly the same amount (37 percent) was found in highly dispersed clusters with less than 5 million square feet of space.

Spatially, complexity is witnessed in the widening mismatch between the geography of commuting (many-to-many) and the geometry of traditional transportation infrastructure (radially focused on the CBD, a legacy of the anachronistic monocentric city). Suburban gridlock is today being eclipsed by exurban and even semi-rural gridlock. Between 1990 and 2000, mean commute time rose 14 percent, to 25.5 minutes, casting doubt over the co-location theory (that holds workers and firms adjust their locations to maintain a constant average commute time). Barriers to mobility, like imperfect information, social exclusion, and large-lot zoning, continue to drive a wedge in the widening spatial mismatch of residences and job locations.

The equity implications of these trends continue to be magnified as America's neediest populations remained concentrated in and around core cities while job opportunities (especially in low-skilled occupations) are mainly in the suburbs and beyond. Many low-income inner-city residents face reverse-commutes via public transit to reach not only jobs but also job interviews, child-care services,

---

[8] R. Lang, *Edgeless City: Exploring the Elusive Metropolis.* Washington, D.C.: Brookings Institution Press, 2002.

and evening adult education facilities. The complexity of travel is seen in the reverse-commute origin-destination patterns of work trips by inner-city residents of Los Angeles. Figure 9.1 shows the location of low-income jobs and households and Figure 9.2 reveals the spatial distribution of reverse commutes (based on 2000 journey-to-work census data).[9] The typical low-income Los Angeles worker who must reverse-commute takes just over an hour to reach his or her job by public transit, roughly twice as long as the average Southern California commuter. Because low-income workers often have contract, contingent-labour jobs, many end up working non-traditional schedules, such as on-weekends and late-shifts, periods when public transit is sparse or non-existent. The complexity of travel patterns over space and time calls for non-traditional forms of mobility that mimic the service characteristics of the private car. This is borne out by research showing that access to a private car better explains successful welfare-to-work transitions among low-income workers in California than quality of public transit services.[10] Clearly, travel complexity raises huge social and environmental justice issues, such as how to best deploy transit. Indeed, the Los Angeles County Metropolitan Transit Authority (MTA) was reprimanded by a circuit judge for concentrating its investment program on high-speed rail systems that would mainly benefit professional-class suburbanites at the expense of inner-city bus users. A consent decree mandates that the agency redirect spending to beef-up traditional bus services, though little progress has been made so far in opening up the marketplace to paratransit competition.

Connections

Of course, the powerful megatrends outlined above do not stand in isolation but rather are usually reinforcing and cross-nurturing. Global competition in a consumer-oriented society has prodded more and more firms to seek out isolated, secluded locations, be it to

---

[9] R. Cervero, et al., Reverse Commuting in California: Challenges and Policy Prospects. Sacramento: California Department of Transportation, 2002.

[10] R. Cervero, O. Sandoval, and J. Landis, "Transportation as a Stimulus to Welfare-to-Work: Private Versus Public Mobility", *Journal of Planning Education and Research,* Vol. 22, 2002, pp. 50-63.

protect one's business culture, trade secrets, or top talent from corporate raids – thus giving rise to the "edgeless" city. The resulting scatteration has compelled automobile-dependent living most prominently among those least able or willing to walk, bike, or endure the discomforts of public transport usage – such as seniors, one of the fast- growing segments of America's population. Scatteration, coupled with womens' massive labour-force entry over the past two decades and the challenges they face in juggling professional and child-rearing responsibilities, has spawned zig-zag travel patterns that are virtually impossible to serve by any form of mobility than the private car. Land-use segregation has spawn chained trip-making, equally car-dependent in nature.

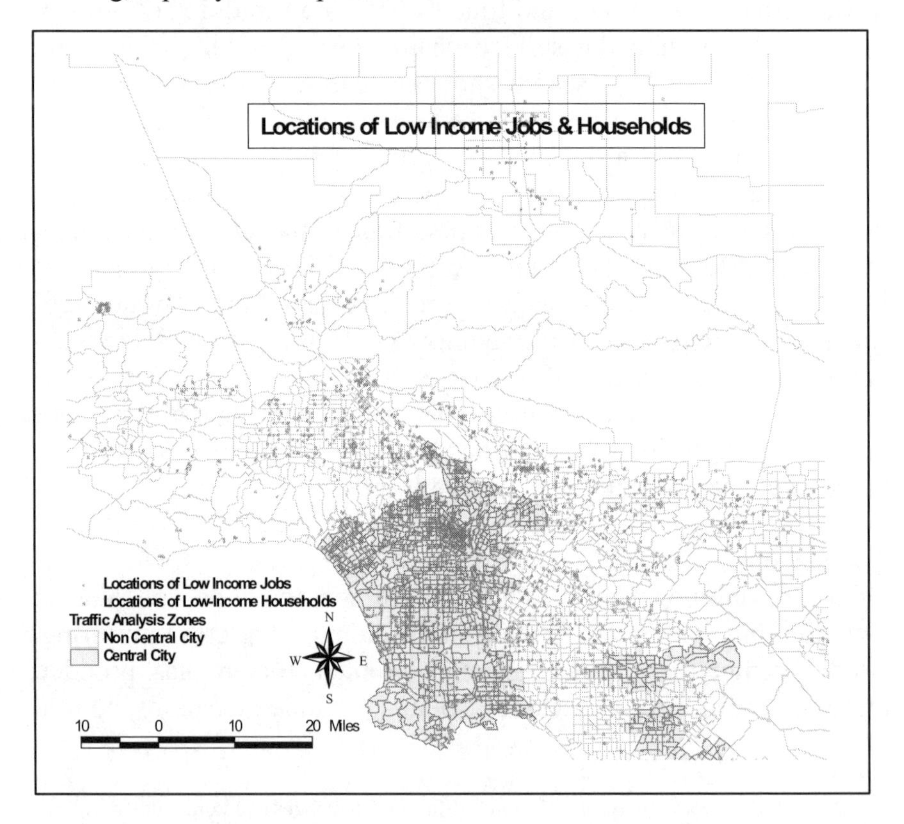

Figure 9.1. Location of Low Income Jobs & Households in Los Angeles County, 2000

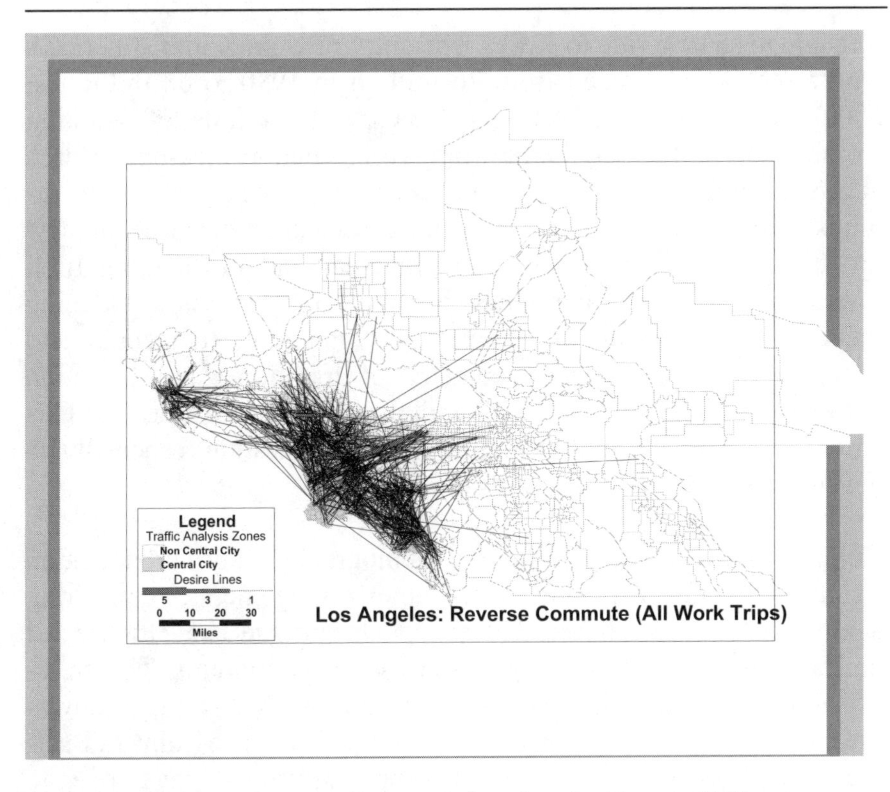

Figure 9.2. Reverse-Commute Patterns in Los Angeles County, 2000

Transportation statistics reveal the mobility implications of this confluence of events. Despite investing tens of millions of dollars in high occupancy lane (HOV) facilities over the 1990s, carpooling's market share of commutes has been steadily eroding. The share of commuters pooling to work declined from a nationwide average of 13 percent in 1990 to 11.4 percent in 2000.[11] In metropolitan Washington, D.C., traditionally one of America's strongest vanpooling markets, ridesharing has steadily fallen particularly rapidly over the past decade, rooted in the shift from predominantly government to increasingly high-technology employment. Many of the region's software engineers and Internet-industry workers keep irregular hours and rely on their cars during the midday, making it nearly im-

[11] R. Poole and C. K. Orski, HOT Networks: A New Plan for Congestion Relief and Better Transit, RPPI Policy Study 305, February 2003 (http://www.rppi.org/ps305.pdf)

possible to share a ride to work. The entry of women into America's workforce, which soared from 26 million in 1980 to 68 million in 2000, has fuelled trip-chaining – nearly two-thirds of working women stop on the way home from work, often to pick-up children at day-care centres.[12] Telecommunication advances continue to diminish the need for spatial proximity, hastening the pace of new growth on the edges of metropolitan areas and in far-flung rural townships. As growth continues to spread out, there is a widening mismatch between the geography of commuting (tangential and suburb-to-suburb) and the geometry of traditional transportation networks, which tend to be of a radial, hub-and-spoke design. Circuitous trip patterns and mounting traffic congestion, especially in the suburbs and exurbs, have resulted.

Collectively, evolving economic, demographic, and urbanization trends have formed new space-time arrangements, conspiring against all forms of movement except the private car. Figure 9.3 portrays this evolution along a space-time continuum. The traditional monocentric city with concentrated activities (e.g., downtowns and 8-to-5 work schedules) supported point-to-point rail services reasonably well. As technology advances gave rise to polycentric settlements and less regular time schedules, more flexible forms of collective-passenger transport, like bus transit and carpools, prospered. As cities and the regions of the future become increasing "non-centric" and time schedules less certain and predictable, the frontier of space-time possibilities has expanded considerably. An immense challenge faced by the U.S. transportation decision-makers is how to pursue the balanced agenda of mobility, accessibility, sustainability, and livability in light of these protracted, complicating trends.

[12] S. Sarmiento, Household, Gender, and Travel. Washington, D.C.: Federal Highway Administration, 1998.

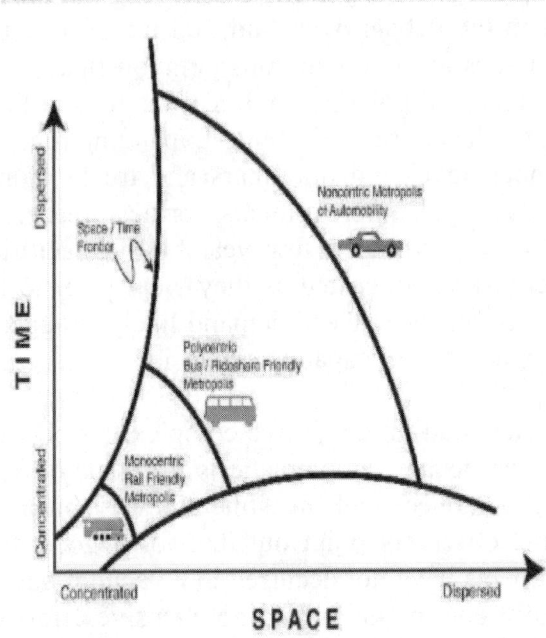

Figure 9.3. Space-Time Diagram of Contemporary U.S. Mobility Demands

Technocratic Complexity

A complex mobility marketplace underscores the immense challenges in advancing knowledge that objectively and faithfully informs public-policy choices. Quite often, different conceptual frameworks, operational assumptions, methodologies, analytical styles, and modes of interpretation have given rise to sharply contrasting empirical insights. Conflicting research findings, contrasting policy interpretations, statistical malaise, and often-times "paralysis by analysis" have all too often been by-products.

Collectively, these analytical dilemmas represent a form of technocratic complexity. Analytical cross-wiring has been poignantly played out in the debate over "induced travel demand". Few contemporary issues in the urban transportation field have elicited such strong reactions and polarized political factions in the United States as claims of induced travel demand. Expanding road capacity is said to spawn new travel and draw cars and trucks from other routes. Consequently, road improvements, critics charge, provide only ephemeral relief—within a few year's time, facilities are back to square one, just as congested as they were prior to the investment. Failure to account for induced demand likely exaggerates the travel-time savings benefits of capacity expansion.

Methodological and interpretative complexities pervade the induced demand policy debate. Emblematic is the issue of causality—might traffic growth induce road investments every bit as much as vice-versa? Some observers point out that for a good century or more road investments have not occurred in a vacuum but rather as a consequence of a continuing and comprehensive effort to forecast and anticipate future travel demand. Accordingly, road improvements act as a lead factor in shaping and a lag factor in responding to travel demand. A study by the Urban Transportation Centre at the University of Illinois at Chicago lends anecdotal credence to this position. Using 60 years of data, the study showed that road investments in metropolitan Chicago could be better explained by population growth rates a decade earlier than vice-versa.[13] For both the Tri-state Tollway (I-294) and East-West Tollway (I-88), the researchers concluded "major population gains occurred in proximity to the expressways over a decade before the construction of the respective expressways".

Failure to account for two-way causality between road investments and highway demand has likely led to inflated claims of the induced demand phenomenon and, as a result, distorted highway investment policy. Many other methodological dilemmas faced in studying induced demand – resolution, measurement, and specification – thwart

[13] Urban Transportation Centre, Highways and Urban Decentralization, Chicago, University of Illinois at Chicago, Urban Transportation Centre, Research Report, 1999.

research progress. Road improvements reverberate throughout a road network, including facilities connecting to an enhanced segment. Tracking the source and geographic scope of new demand is exceedingly complex. Conceptually, Figure 9.4 presents a normative framework for gauging induced demand impacts. The causal chain works as follows: a road investment increases travel speeds and reduces travel times (and sometimes yields other benefits like less stressful driving conditions, on-time arrival, etc.); increased utility, or a lowering of "generalized cost", in turn stimulates travel, made up of multiple components, including new motorized trips (e.g., latent demand previously suppressed), redistributions (modal, route, and time-of-day shifts), and over the longer term, more deeply rooted structural shifts like land-use adjustments and increased vehicle ownership rates (that in turn increase trip lengths and vehicle miles travelled, (VMT). Some of the added trips are new, or induced, and some are diverted. While evidence on the induced-growth effects of new highways is limited, roads and prominent fixtures of America's suburban landscape -- big-box retail, edge cities, and campus-style executive parks – that they serve are clearly co-dependent.

This normative framework was adopted in several recent studies of induced-demand in California. In one, a path-model framework was used to sort out "induced demand", "induced growth", and "induced investment" effects.[14] Recorded traffic increases along expanded freeways were explained in terms of both faster speeds and land-use shifts. Because less than half of the recorded speed increases were statistically attributable to road improvements, a fairly modest long-term induced-demand elasticity of 0.39 was recorded. The longitudinal effects of rising VMT on roadway investments were of a similar order of magnitude. This path analysis produced elasticity estimates considerably below those of earlier studies (that have generally been in the 0.7 to 0.9 range),

[14] R. Cervero, 2002, "Road expansion, urban growth, and induced travel: A path analysis", *Journal of the American Planning Association*, Vol. 69, No. 2, 2003, pp. 145-163.

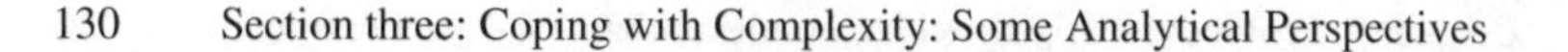

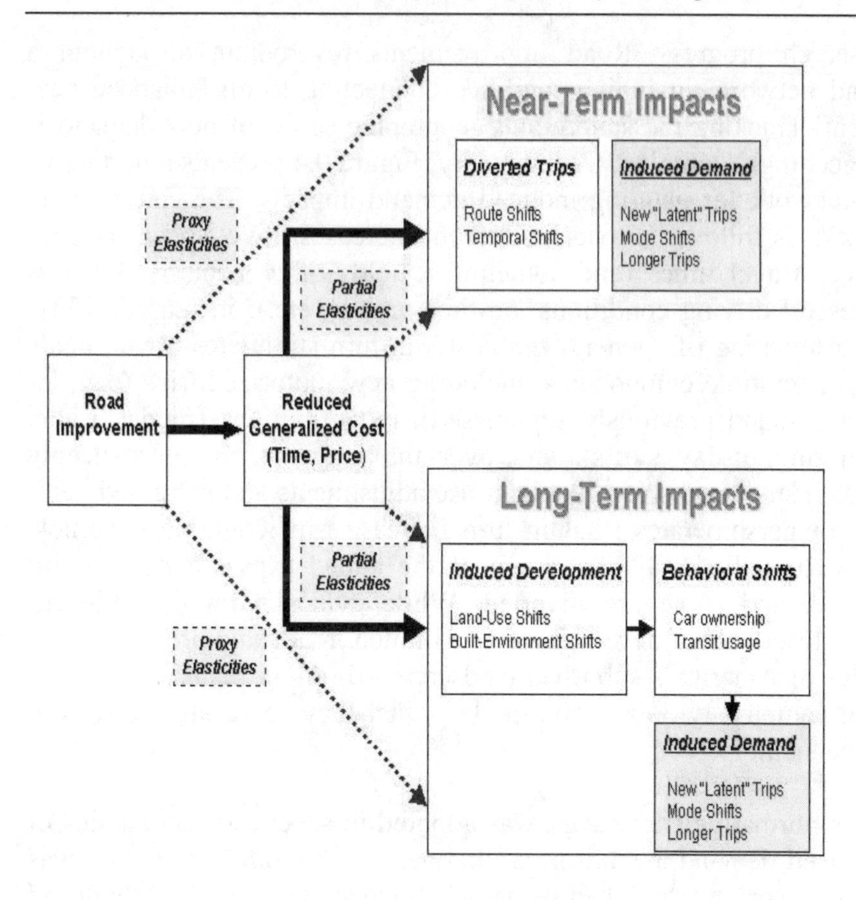

Figure 9.4. Normative Framework for Studying Induced Travel Demand

underscoring the fact that dramatically different results can be produced under different model specifications. Overall, models that have sought to account for two-way causality have yielded lower elasticity estimates (in absolute terms) than those based on simpler, single-equation analyses.

Going from hypo-deductive research to operational transportation modelling and forecasting adds more layers of complexity. Douglas Hunt portrayed the many ways in which second-order induced-demand impacts need to be accounted for within the framework of traditional four-step travel-demand forecasting models.[15] Figure 9.5 portrays the full array of spatial activities and relationships encapsulated in traditional four-step models. Figure 9.6 reflects the pathways in which induced demand – as reflected by route, time-of-day, modal, and land-use shifts – need to be accounted for within the traditional long-range demand-forecasting modelling framework. In practice, lack of empirically demonstrated relationships renders operationalizing such adjustments intractable. Hunt et al. made some headway in their modelling and forecasting of induced-demand effects in investigating various transportation and land-use scenarios for metropolitan Sacramento.[16] Models revealed appreciable induced-demand impacts attributed to locational shifts, reflected by feedback mechanisms in dynamic models (with large variations depending upon the specific modelling platform – MEPLAN, TRANUS, DRAM/EMPAL, or SACMET – that was used). For most regional planning entities within the United States, model platforms are nowhere near sophisticated or robust enough to incorporate such complexities. The inability to account for induced-demand within formal modelling frameworks means that, to some degree, road investments unavoidably become more political in nature than they otherwise would be.

[15] D. Hunt, "Induced Demand in Transportation Demand Models", *Working Together to Address Induced Demand*, Washington, D.C., Eno Transportation Foundation, 2002.

[16] D. Hunt, et al., "Comparison from Sacramento Model Test Bed", *Transportation Research Record 1780*, 2001, pp. 53-63.

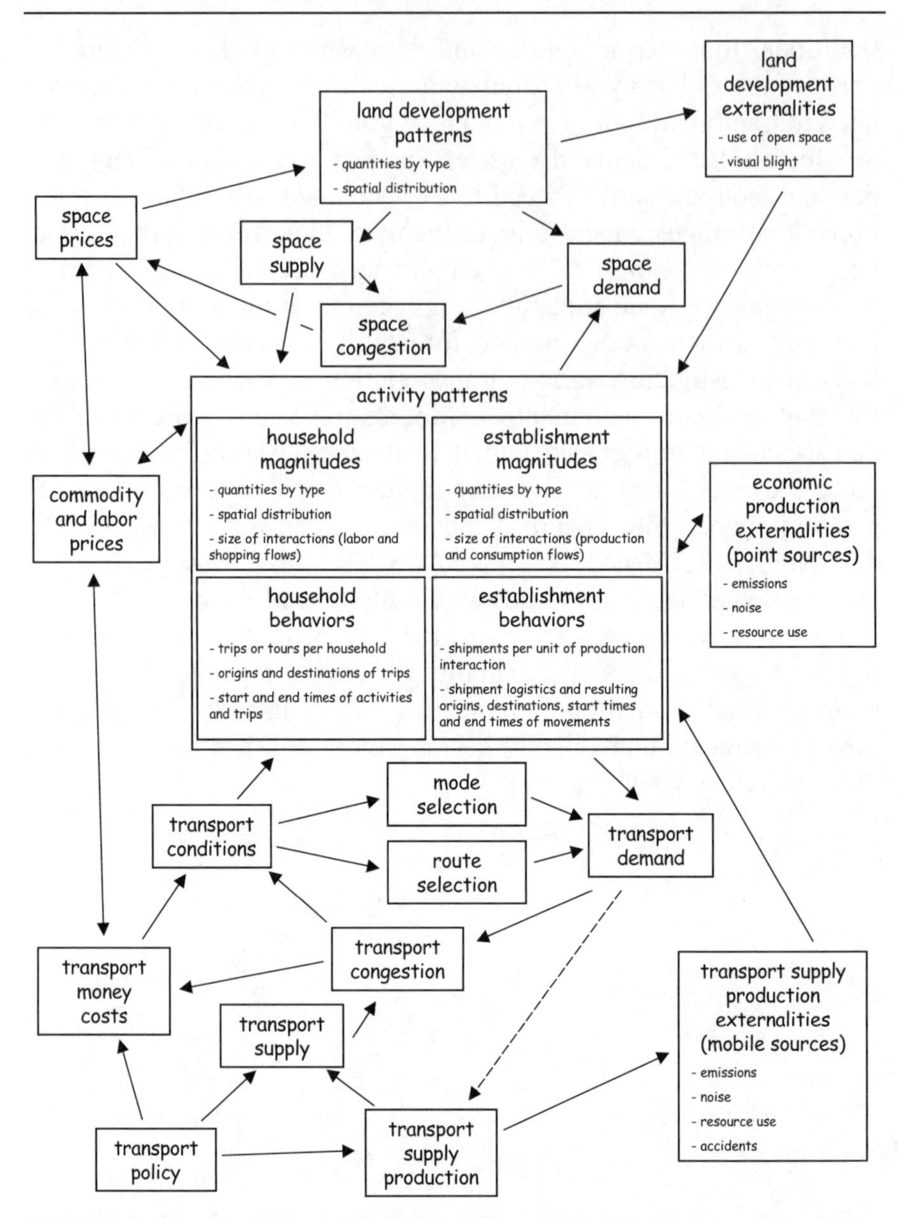

Figure 9.5. Systems Framework for Modelling Spatial Activities and Travel Demand.
Source: D. Hunt, Induced Demand in Transportation Demand Models, *Working Together to Address Induced Demand*, Washington, D.C., Eno Transportation Foundation, 2002.

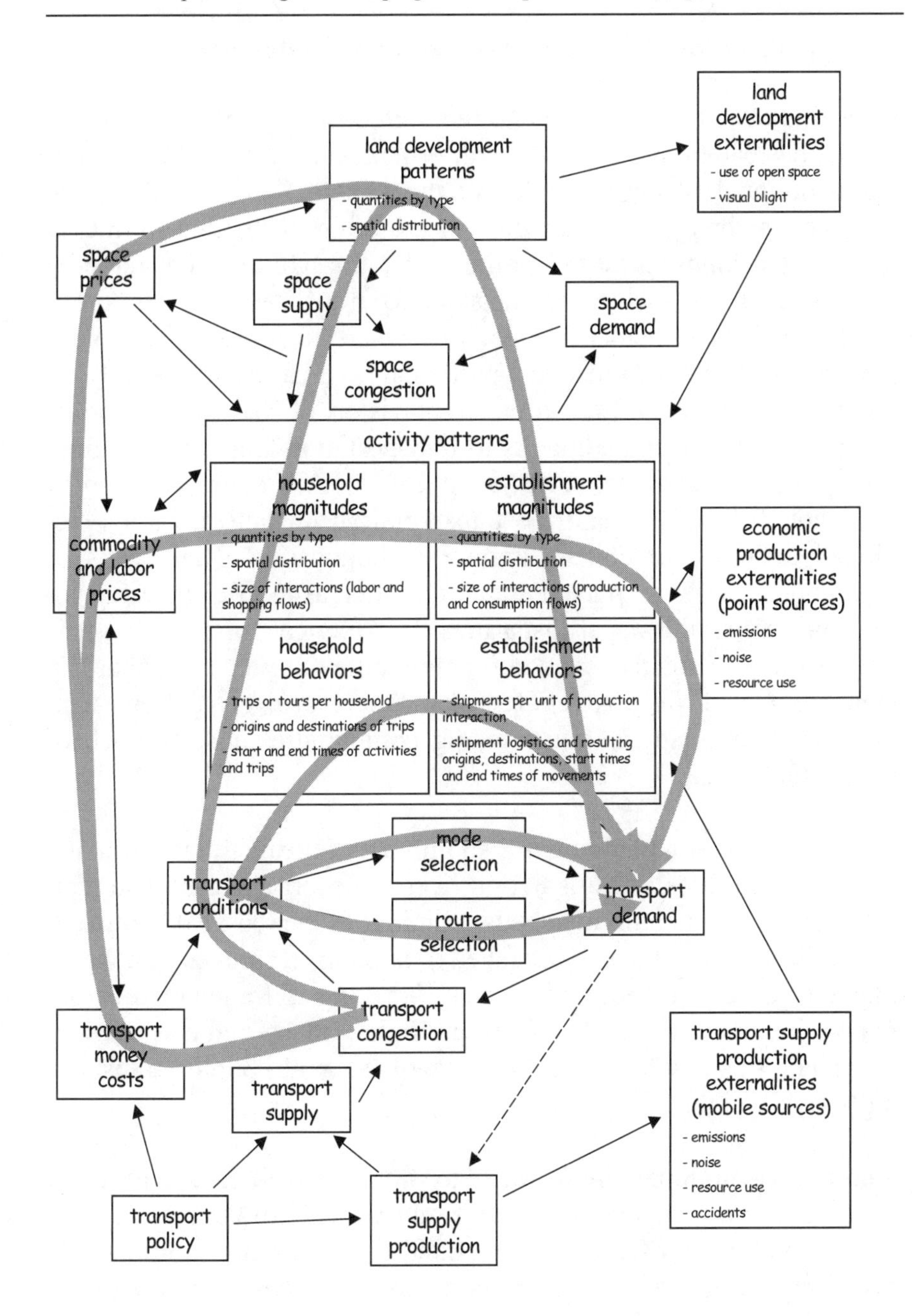

Figure 9.6. Pathways for Incorporating Induced Demand Effects Within Modelling Framework. The same source as in figure 9.5.

Complexity in the Political and Institutional Landscape

Concerns over sustainability and the high economic costs of serving sprawl has catapulted smart-growth principles to centre stage within many regional planning circles of the United States. In America, however, pathways to smart growth are often obstructed by messy institutional landscapes and political detours. Quite often, regional land-use patterns – which set the stage for travel – are the sum product of local, incremental decisions on where to locate a new shopping plaza, whether to rezone a particular land parcel, etc. Rarely do these decisions shape into a coherent vision of the future. One of many institutional impediments to transportation-land use coordination is the mismatch between where decisions on land development are made -- locally -- and the transportation impacts are felt -- regionally. Travel, of course, knows no boundaries. The effects of poor coordination get played out all too often as inefficiencies, negative spillovers, and fiscal disparities. In America, for instance, it is not uncommon for fast-growing communities to place regional trip generators, like big box retailers that fatten local tax coffers, near their boundaries so that surrounding communities absorb much of the traffic burden.

U.S. transportation planning is also mired by bureaucratic inertia and redundancies. Ideally, jurisdiction over transport and land-use matters would match commutersheds -- similar to the regional context in which water resources (watersheds) and air resources (airsheds) planning occurs. In practice, decision-making is fragmented across many jurisdictions and often multiple transportation service-providers (e.g., separate entities involved with public transport, highways, freight, ferry services, etc.).

Another institutional impediment to smart growth is the irregular pace of land-use change. Local and subregional growth often occurs incrementally, in fits and starts. Land-use maps are continuously changing because of zoning amendments, variances, and new subdivisions. In contrast, decisions on regionally important transportation improvements often occur in 2 to 3 year time increments, and are hard to reverse or change in response to unfolding land-use patterns.

Thus whereas land use changes are fluid and on-going, large-scale transportation projects tend to be rigid and occur over much longer time increments.

Also hampering coordination is the reality that the benefits of careful transport-land use integration are often not evident until ten or more years in the future. This is inherently at odds with political systems that demand short-term payments, IMTO ("in my term of office"). Elected officials are much more likely to embrace a large-scale road project that immediately relieves congestion and generates lots of jobs and political capital than transit villages, jobs-housing balance, New Urbanism, and other land-use strategies with questionable near-term pay-offs.

An additional institutional impediment is the difficulty in forging any degree of consensus or vision on desirable land-use futures in a highly pluralistic, freely democratic society like the United States. This is magnified by smart-growth initiatives, however well intended. Smart growth planning and development embraces the principle that an overarching vision should guide the integrated and sustainable transportation planning process, reflecting the fact that travel is fundamentally a "derived demand" – derived by the need to get to and from places or activities. In this sense, transportation is a means to the "land use" end of a trip. Since land use speaks directly to activities that take place over space, normative planning calls for land-use visions to take precedence over transportation visions (with the understanding that transport infrastructure can be a powerful tool for shaping land-use visions). Great examples of cogent land-use visions (based on sustainable urbanism principles) that guided transport investments include Copenhagen (Finger Plan), Stockholm (Planetary Plan), Curitiba (Linear City Plan), and Bogota (Egalitarian City). These communities profited from the presence of visionaries, like Sven Markelius, Jaimie Lerner, and Enrique Penalosa, who could elegantly articulate visions and rally broad-based political support for their visions. In the United States, increasing socio-ethnic diversity and the constitutional protections governing individual freedoms (including personal property rights) means that building any degree of consensus on what constitutes a desirable future is next to impossible. In a pluralist society like the United States, opinions and preferences are scattered all over the map regarding the desirability of compact, mixed-use, pedestrian-friendly development. This all too often has been manifested by U.S. cities aggressively moving forward with light-rail transit investments without any serious thought about the kinds of built environments necessary to sustain these costly outlays. America is littered with examples of clear transportation visions (e.g., modern point-to-point rail systems) absent an articulated land-use vision. This has often meant land development turning its back on rail transit—e.g., lots of campus-style office parks, mega-malls, and big-box retailers and the designation of park-and-ride lots as the dominant land use around rail stops.

Promising Developments for Coping with Complexity in America's Transport Sector

Notwithstanding the complicating effects of megatrends and technocratic-institutional roadblocks, progress is being made in parts of the United States to respond to unfolding market trends, integrate transport and land-use, and institute new institutional arrangements promote efficiency, accountability, and social justice. This section reviews some of these experiences.

Responding to Mobility Markets

Two U.S. examples of responding to emerging mobility markets are car-sharing and client-based mobility initiatives for needy workers. While carsharing has been around for several decades in Europe, only over the past few years has it gained a foothold in American cities, most notably Boston, Seattle, and San Francisco. Besides having fairly high densities and mixed land-use characters, perhaps what these cities most have in common with their European counterparts are limited and expensive car parking. In a recent evaluation of San Francisco's City CarShare program, I found that vehicle miles travelled (VMT), adjusted for vehicle occupancies and engine-size of automobile trips, generally rose faster for those who joined

City CarShare than a control group.[17] Given that around two-thirds of surveyed City CarShare members come from zero-car household, the sudden availability of cars likely stimulated automobile travel for some. Motorized travel appeared to replace some trips previously made by foot or bicycle. Presumably carshare trips have high value-added in that members pay market-rate prices for use of cars. The majority of carshare trips did not correspond to the peak periods, suggesting many carshare trips did not contribute to traffic congestion. "Judicious automobility" should be looked upon in a positive light since travel desires are being met while keeping the population of private cars lower than it otherwise would be.

As carsharing matures and its membership becomes more mainstream, travel-behaviour impacts appear to be changing with time. Evidence from a second-year survey suggests this is indeed the case. Over 70 percent of members had gotten rid of a car or forwent the purchase of a new car by year-two. Also, travel-diary data suggests VMT per capita went down faster for carshare members almost twice as fast as for a control group (for weekday, workday travel). These findings suggest the availability of a shared car has spurred significant numbers of San Francisco households to get rid of a second car within 24 months, and that this in turn spurred more efficient travel and perhaps occasional foot and bicycle trips for in-neighbourhood convenience shopping. These results suggest that as new members are drawn from the ranks of car-owning households, the relinquishment of private cars will eventually suppress motorized travel. Innovative, market-oriented car-based strategies like carsharing (along with station cars) hold considerable promise in good part because they respond to shifting demographic and urbanization trends.

In response to welfare-to-work concerns, some U.S. cities and regions have aggressively pursued client-based strategies. A leader in this arena has been San Cruz County, California. There, social-service professionals work with newly employed individuals re-

[17] R. Cervero, "City CarShare: First Year Travel-Demand Impacts", *Transportation Research Record*, 2003.

cently weaned from welfare to custom-design mobility programs tailored to their particular commuting needs. A host of options are available including fairly expensive door-to-door shuttle service (for outlying areas poorly served by public transit), emergency rides home, carpool incentives, work-related emergency payments, mileage reimbursement, and bus passes. The shuttle program not only *connects* needy people to jobs, but also *creates* jobs. Notably, welfare-recipients are trained and hired to drive vans, enabling them to gain firsthand experience in the van business. Evidence suggests custom-tailored programs, which costly on a per capita basis, better achieve hoped-for outcomes than provider-side, transit-based programs – namely, greater success at inducing welfare-to-work transitions.

Also successful has been the family loan program, practiced in Santa Cruz and several other northern California counties, that provides small loans to welfare recipients and low-income parents for purchasing cars. The loans are serviced by four local banking partners that are able to access low-interest federal funds under the Community Reinvestment Act (CRA). Through car ownership, clients are getting to work more quickly and on-time: 18 months into the program, loan recipients reported a 93 percent average reduction in time spent getting to work and a 90 percent decline in work time missed. Additionally, there was a 26 percent increase in attendance at job-related educational activities. Perhaps of most importance are "outcome" measures – i.e., to what degree did the loans achieve their intended purpose of promoting welfare-to-work? The best indicator is that average gross incomes rose after loans were issued: by 23.8 percent within the first 6 months of receiving a loan and by 36.9 percent at the end of the loan term.

There are other signs of adaptability to changing mobility markets, such as the launching of station cars and operation of laissez-faire paratransit (from microbus-jitneys in Miami to shared-ride taxis in Berkeley to pedicabs in Manhattan). A market-oriented initiative on the highway side has been high-occupancy toll (HOT) lanes. HOT lanes have been in operation for several years in Orange County (the SR91 Express Lanes) and San Diego County (converted HOV lanes on I-15). A 2001 survey found 91 percent of motorists travelling the I-15 HOT-lane corridor supported choice afforded by HOT lanes.

TOD and Adaptive Re-Use

Transit-oriented development (TOD) has gained currency as a means of curbing sprawl, reducing traffic congestion, and expanding housing choices. Research underscores the mobility and land-development benefits of TOD: if well designed, concentrated, mixed-use development around transit nodes can create a ridership bonus of 200 to 300 percent (above comparable development away from transit) and a land-price value-added as high as 100 percent.[18] Ridership gains, research shows, are significantly a product of self-selection, with those with a lifestyle predisposition for transit-oriented living conscientiously sorting themselves in apartments, townhomes, and single-family units within an easy walk of a transit node. For the San Francisco Bay Area, nested logic modelling revealed that upwards of 40 percent of the ridership bonus associated with TOD is a product residential self-selection. This finding underscores the importance of introducing market-responsive zoning in and around transit nodes – zoning that acknowledges that those living near transit tend to be in smaller households with fewer cars. Flexible parking standards are one initiative introduced in some U.S. settings. Also promising are Location Efficient Mortgage (LEM) programs that make it easier for someone to purchase a home near transit stations (reflecting the fact they will likely spend less money on automobility as a result).

Currently, over 100 examples of TOD exist in the United States. Most impressive has been the Rosslyn-Ballston corridor in Arlington County, Virginia – since Washington Metrorail's opening in the late 1970s, some 14 millions square feet of office space and 25,000 housing units have been built within a quarter-mile of rail stations. Seven times the land area would have been necessary to accommodate this growth (which today accounts for 52 percent of the

[18] R. Cervero, C. Ferrell, S. Murphy. Transit-Oriented Development and Joint Development in the United States: A Literature Review, Research Results Digest No. 52, Transit Cooperative Research Program, 2002.

county's tax base) if built at suburban standards. Portland, Oregon is another successful U.S. example of TOD, a product of several decades of revitalizing the downtown, ramping up transit services, and targeting infill housing development (and master-planned projects, like Orenco) to station areas (see chapter 14). Portland's share of work trips by mass transit rose18 percent during the 1990s, bucking a trend toward declining transit market shares in many other parts of the United States. While some critics charge urban containment policies and TOD have increased housing prices, most serious studies of the situation suggest demand to be in a well-planned and highly liveable U.S. city like Portland explain rising prices more than restricted land supplies.[19]

One of the more efficient land-use changes occurring in the United States has been the adaptive re-use of superfluous surface parking lots at transit stations. Car parks are proving to be a blessing in disguise for they provide large swaths of conveniently located, preassembled land with great regional accessibility. Most attractive are surface parking lots at train stations. Many were originally overbuilt, thanks to generous federal funding for rail development. As areas have matured and surrounding land values have increased, market pressures are prompting U.S. transit agencies to sell off at least portions of them as a means to both create a ridership base and to reap windfalls in the form of value capture. Often, the profits earned are more than enough to cover the cost of replacement structured parking, freeing up land for infill development. Surface parking conversion, then, is a back-door form of land-banking, which in many European cities, including Stockholm, has been a principle means of leveraging transit-oriented development.

The city of San Jose, California and the Santa Clara Valley Transportation Authority (SCVTA) recently joined forces in designing a mid-rise, mixed-use project on the park-and-ride lot at the Ohlone-Chynoweth light rail station. Historically, the region's light-rail system has struggled to build a ridership base in large part because

[19] A.Nelson, R.Pendall, C. Dawkins, and G.Knaap, The Link Between Growth Management and Housing Affordability: The Academic Evidence. Washington: The Brookings Institution Centre on Urban and Metropolitan Policy, Discussion Paper, 2002

much of its service territory is the Silicon Valley, a landscape of sprawling office campuses and car-oriented shopping plazas. However, as the demand for affordable housing with good access to the Silicon Valley has intensified, local policy-makers have come to the realization that parking-lot infilling was too good of an opportunity to pass up. At the time of project development, only 30 percent of the 1,140 original parking spaces at the Ohlone-Chynoweth station were used. Already, 500 parking spaces have been converted to 195 units of two and three story town homes, a retail plaza, a child-care facility, and a community recreation centre.

Another promising area is to smartly re-use antiquated and dysfunctional shopping centres. The trend in retailing toward warehouse-shopping, e-commerce, and mega-entertainment malls has led to the closure of many out-dated 1960s and 1970s shopping centres across the United States. Like rail parking lots, one of the biggest assets of dying shopping centres is their huge amount of pre-assembled real estate. One of the more successful adaptive re-uses of a shopping centre and integration with rail transit is The Crossings project in Mountain View, California. The Crossings is an 18-acre compact, mixed-use, and walkable neighbourhood near a commuter rail line some 30 miles south of San Francisco. It replaced a slowing dying shopping centre and movie theatre that were surrounded, in big-box fashion, by a huge, underutilized surface parking lot. The project's 540 housing units have commanded a rent premium, partly because of proximity to rail and partly because of the high-quality of urban design. Many well-paid young professionals with jobs in downtown San Francisco and the nearby Silicon Valley have opted to buy into The Crossings, drawn by its ambience and exceptional accessibility to transit. Generous landscaping and public spaces punctuated by an internal pathway network have created a highly attractive urban milieu, notwithstanding residential densities of 30 units per acre, fairly high by suburban California standards. Zero-lot lines and rear-lot parking have allowed such densities to be achieved. As a gateway to the Mount View CalTrain station, The Crossings stands as one of the few transit villages oriented toward commuter rail.

New Institutional Arrangements

As noted earlier, traffic congestion in much of the U.S. stems, in part, from ineffective institutional structures that lead to a discordance between regional land-use and growth-management planning and regional transportation investments. While some states like Florida and Maryland have made progress in advancing concurrency laws that mandate land-use and transportation infrastructure be harmonized, for the most part ineffective institutional arrangements have resulted in mismatches between urbanization and infrastructure development. The state of Georgia has made a bold departure in this regard by forming an all-powerful regional transportation authority that is well-positioned (with purse-string powers at its side) to coordinate mobility planning and land-use development. Called the Georgia Regional Transportation Authority (GRTA), the organization not only oversees the planning and expenditure of funds for all urban transportation improvements in the state, but also has broad control over regionally important land uses, like shopping malls, industrial parks, and sport stadia. Local land-use decisions must conform to broader regional transportation and development goals, otherwise GRTA can effectively veto the decision by threatening to cut off all state infrastructure funds. GRTA's formation was largely in reaction to decades of poorly planned growth in metropolitan Atlanta, matched by ever-worsening traffic congestion. The announced plan of a large high-technology employer to relocate out of Atlanta because of unsustainable traffic congestion and a declining quality of life was a political wake-up call. The region's new planning philosophy — one of balancing urbanisation and transportation investments — aims to enhance mobility while also placing the region on a smart-growth pathway. The ability of GRTA to leverage the mix-use transformation of an in-city brownfield site abandoned by the Atlantic Steel company into a mixed-use village has been an important victory for smart growth. For purposes of securing federal infrastructure funds currently frozen because of Atlanta's violation of air quality mandates, GRTA and others successfully argued that infill development would be less harmful to Atlanta's air basin than comparable growth on the car-dependent edges.

Close: Expanding Choices in a Complex World

While complexity in America's mobility marketplace and institutional landscape has stymied efforts to move forward with bold transportation initiatives, it has also given rise to a mindset that calls for more flexibility, market-responsiveness, and variety in the transportation/land-use arena. Although critics of smart-growth planning equate it with social engineering, in truth anything that widens choices in where to live, work, and shop as well as how to travel is inherently in society's best interest. Clearly, living in compact, mixed-use, easily walkable communities is not for everyone. Middle-class and well-to-do households with several or more children and a preference for privacy and seclusion will continue to reside mostly in the suburbs and beyond. Back-office functions will continue to flock to outlying and far-flung places where real estate prices are cheaper. Big-box retailers and multiplex cinemas will continue sprouting on the outskirts. Smart-growth initiatives in no way intervene in such free-market locational choices as long as those making the choice pay something which comes reasonably close to reflecting true social costs. Rather, smart growth – whether in the form of an infill housing project on a former transit parking lot or an edge city with a balance of jobs-to-housing and roads-to-busways – is mainly about expanding choices and offerings in a free market context. More variety in housing choices, in particular, is an adaptation to the steady growth in single-person households, childless couples, and empty-nesters, many of which prefer in-city, small-lot living in attractive environments that are well-served by public transport and easy to get around by bike and foot. Variety and choice is something that finds broad political and ideological appeal. It is precisely for this reason that integrated transport and urbanism – despite the many barriers that must be overcome -- is likely to prevail as America's dominant paradigm of community-building in the twenty-first century.

Chapter 10. Self-organizing Systems and Innovations in Asian Cities

Geetam Tiwari

Urbanization patterns

A large proportion of the urban population in Asian and other low income cities remains outside the formal planning process. Survival compulsions force them to evolve as self organised systems. These systems rest on the innovative skills of people struggling to survive in a hostile environment and meet their mobility and accessibility needs. The actors in this complex street environment cannot be wished away. They are here to stay.

The informal sector in Asian cities is an example of self-organizing system which can provide us with useful insights to address future urban transport problems. This sector continues to be viewed as an 'unwanted' sector in the city, and hence formal plans for housing and transport do not have provisions for their needs. Yet this sector exists in every megacity, and millions of people across Asia survive because of the innovations taking place in this self-organizing system.

Asia is home to some of the most populous cities in the world. Even the future growth of cities is expected to occur in Asia because most countries have less than 50% urbanization at present. Undoubtedly what happens in this region has immense global implications. Asian cites which a have high population density continues to experience rural-urban migration unlike Latin American cities which already have high densities but where the region is already 80% urbanized. The African region has a smaller proportion of population compared to Asia and a low level of urbanization as well as a low population density compared to other regions.

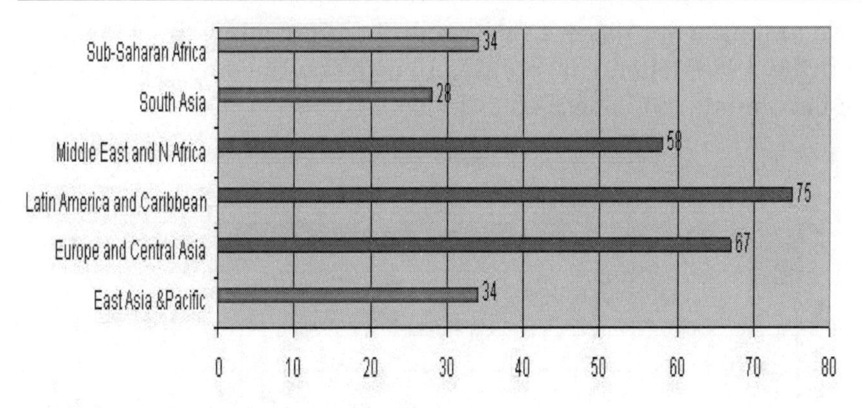

Figure 10.1 Urbanization in world regions

Amongst low and middle income regions of the world, the Asian region, which is already home to 1 billion urban population, has potential for rapid urbanization in the immediate future. Understanding and managing this region is the key to solving Future Urban Transport problems. Let us look at the urbanization trends in Latin America which is already 75% urbanized along with two major regions in Asia, China (around 35%) and India (around 28%), where the urbanization process has ranged from completely controlled to paper plans only .

Unlike Latin America where rapid urbanization occurred between 1940 and 1980 (Figure 10.2), the Asian region experienced slow urbanization. Between 1950 and 1970, Chinese cities followed planned urbanization. In the initial years the number of people living in urban areas was less than the planned population. However, after 1970 the number of people living in urban areas exceeded the number planned for, and this trend continues to this date (Figure 10.3).

All these regions have experienced an increase in heterogeneity, with economic and population growth. The poorer section of the population which remains outside the formal planning process have survived and improved their lives compared to their rural counterparts by constructing shelters on 'invaded land'. In Latin American cities they have been forced to occupy land on the urban periphery, whereas in Indian cities they are often in the centre of the city or close to their work place. Chinese cities did not allow this to happen until the 1980s although, proliferation of low income housing was not uncommon in Chinese cities in the 90s.

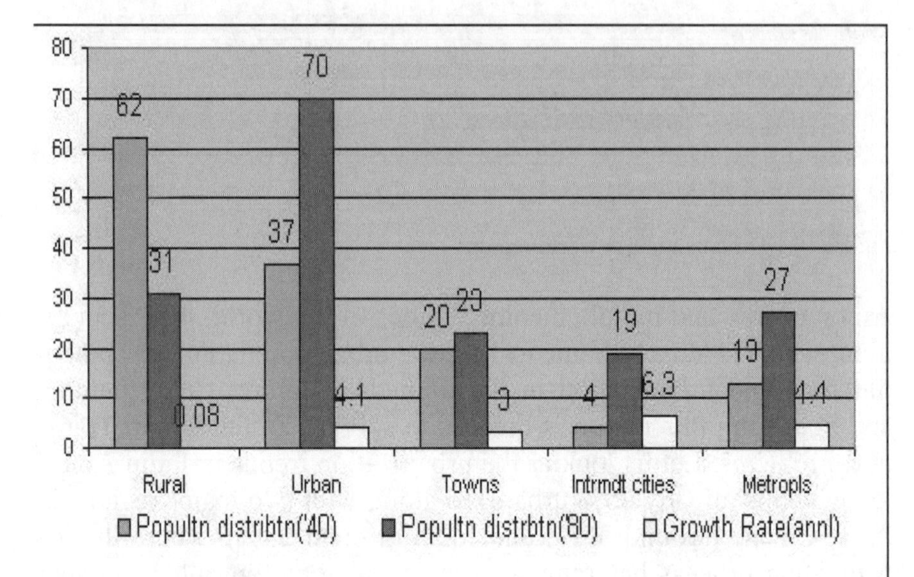

Figure 10.2 Urban growth in Latin America 1930-1990
(80% urbanized in 1990)

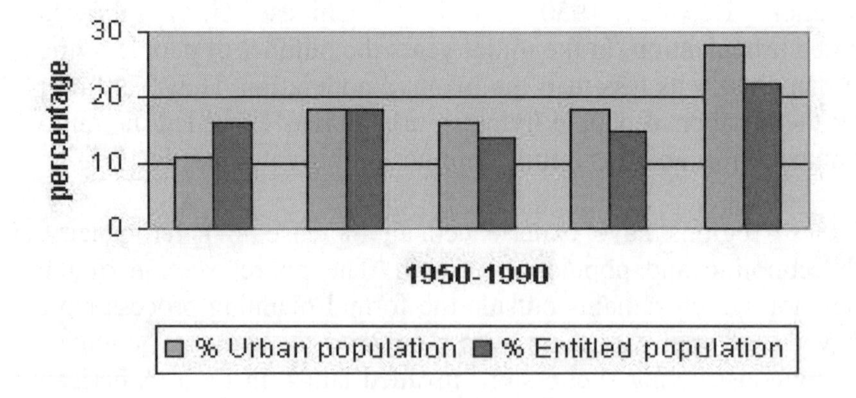

Figure 10.3 Urban Growth in Chinese Cities

Urbanization in India has been relatively slow, yet it is one of the largest urban systems. With the growth in the economy the rate of urbanization is expected to increase, i.e. more people will live in cities. The challenge to meet the future travel demands lies in understanding the character of our cities. These cities are characterized by diversities and heterogeneity in socio-economic conditions. The big cities (Delhi, Kolkata, Mumbai, Chennai) are often agglomerations of several small cities in close proximity to each other with multiple economies. Often 10-50% of the population lives in "slums". The slum population seems to grow as the city size grows. Nearly 30-50% of slum dwellers live in 'unauthorized' self-constructed dwellings, close to work.

City	Total population(million)	Slum population	
Mumbai	12	5.82351	49%
Delhi Municipal Corporation	10	1.854685	19%
Kolkata	5	1.490811	33%
Bangalore	4	0.3452	8%
Chennai	4	0.747936	18%
Ahemdabad	4	0.439843	13%
Hyderabad	3	0.601336	17%
Pune	3	0.531337	21%
Kanpur	3	0.368808	15%
Agra	1	0.12189	10%
Varanasi	1	0.138183	13%
Meerut	1	0.471316	44%

Table 10.1. Slum Population in million plus cities in India

Growth of informal economic activities is common to all three regions. Between 1970-1990 several cities in Latin America achieved social segregation; poor squatters were expelled to make way for new middle class housing and parks, strategies which are being followed in Indian cities in the last five years in the name of making liveable cities. If Latin American cities have 30-50% informal employment, Indian and Chinese cities are not far behind. In all three regions the term 'informal' is synonymous with the term 'bare subsistence'.

By 1990 informal employment grew to 30-50% of total employment in Latin American cities. Poverty is concentrated in the informal sector. Survival strategies include child labour, ignoring mother's nutrition, exclusion from public utilities, abandoning children's education. In Latin American cities household and community survival strategies have led to household and community fragmentation. Often survival strategies in these self-organised systems result in a disregard of community spirit. Already in Latin American cities the reduced possibilities of help has lead to unorganised forms of protests and street crime. Action benefiting individual household interests has weakened community spirit. Therefore, non payment of public utilities is common. Further complexity arises from the fact that the informal economy is not a homogeneous sector. The differences in rural *versus* urban, ethnic minorities and gender inequalities complicate the system. It is crucial to understand that the informal economy and marginal squatter settlements are an integrated part of the urban land market. This is responsible for spatial heterogeneity.

The informal sector

In general, mega cities in low income countries are characterized by diversities and heterogeneity in socio-economic conditions. These mega cities are agglomerations of several small cities in close proximity to each other with multiple economies. One economy serves the needs of the affluent and features modern technologies, formal markets, and the outward appearance of developed countries. The other serves disadvantaged groups and is marked by traditional technologies, informal markets, and moderate to severe levels of economic and political deprivation.

The growth of the informal sector is accompanied by the growth of the formal sector. The informal sector develops firstly because of rural-urban migration, which is based on a comparison of rural and urban opportunities. The formal sector in urban areas requires unskilled labour or a low level of skills provided by migrant labour. Construction labour, casual labour in factories and commercial establishments depend on migrant labour. Higher income families require a domestic help, gardener, carpenter, plumber, electrician, etc.

for normal functioning. Often in Indian cities squatter settlements develop inside the city close to commercial centres, planned housing developments, and factories. Soon these settlements demand all kinds of services needed by the residents - low cost food, vegetables, tailors, etc. It is a common sight in Asian cities to have street vendors along the roads selling food, toys, flowers, and various handicrafts made by family members. Thus the informal sector provides an employment opportunity to each and every family member and rewards them for their creative endeavours. Growth of the formal sector is accompanied by the growth of the informal sector, with the latter showing higher growth rates than in the formal sector. This is not surprising as the informal sector grows to serve the formal sector as well as to serve the informal sector.

The patterns of development in Asian mega cities is an amalgam of planned and organic self-organizing) growth. Most of us see congestion, crowding, poverty and chaos as ubiquitous phenomena, but we fail to recognize the human ingenuity for survival, the social cohesion and low street crime rates present in these cities. If the Asian cities are allowed to follow the pattern of exclusion, we may see sharp rise in street crimes in Asian cities also. The important lesson from these trends is to recognize that the informal economy and marginal settlements are an integral part of cities. Let us plan for it.

Urban travel patterns

These living patterns have major impacts on the urban travel pattern. Urban travel in Asian cities is predominantly walking, cycling and public transport (Table 10.2). The variation in modal split among these three modes seem to have a relation to per capita income, city size and per capita income. Small and medium size cities have a lower income than mega cities, and therefore the dependence on cycle rickshaws and cycles is greater than in larger cities. Shanghai and Beijing, both mega cities with high industrial growth, continue to depend on walking and cycling. In Shanghai, until about 1990, almost all travel was on foot, by bicycle or bus. Cars, scooters, and motorcycles were rare. Travel distances increased in the late 80's

City (Population in million)	Modal split (percentage of daily trips)							
	Walking	Cycles	Public Transport	Two Wheelers	Car	Para transit		
						MTW[20]	CR[21]	See Table 10.3.
Delhi (13)	14	24	33	13	11	1		1, 2, 3,4,5a
Mumbai (14)			88		7	5 (taxi)		1, 2,3,4,5 a
Kanpur (3)	34	18	12	23	0	4	9	1,2,3a, 4,5a
Ahemdabad(5)	40	14	16	24	0	5	0	1,2,3a, 4,5a
Beijing[22] (12)	14	54	24	3	5			1, 2a,3,4 a,
Shanghai[23] (13)	31	33	25	6	5			1,2a,3, 4a
Manila (10)			24+2+3		30	41		2,4
Jakarta (11)	13						12	2,3,5a
Dhaka [24] (14)	62	1	10	4	4	6	13	1,2,3a, 4,5a
Bangkok (7)	16	8	30		46			2,3,4,5 a

Table 10.2. Model split of daily trips, selected cities in Asian countries

[20] Motorised Three Wheeler Taxi

[21] Cycle Rickshaw

[22] Zhou He-Long and Deng Xin-dong, Cycling Promotion and Bicycle Theft, report for Interface for Cycling Expertise, Utrecht, The Netherlands, 1996

[23] Ximing Lu, Xiaoyan Chen, and Xuncu Xu, Urban Transport Planning and Urban Development, Shanghai City Comprehensive Transportation Planning Institute, East China Polytechnic Publishing House, 1996.

[24] Hoque M., Jobir Bin Alam, Strategies for safer and sustainable urban transport in Bangladesh, proc. CODATU X, Lome, 2002.

and 90's, not only due to income growth but also to industrial relocation. Also newly developed low-density areas were too far for bicycles and not profitable for buses, which is why scooters and small motorcycles became a popular mode of travel. Unlike Indian cities, where the scooter and motorcycle are the fastest growing vehicles, the scooter and motorcycle population in Shanghai and other Chinese cities is declining because of new restrictions on the registration of new scooters and other vehicles with two-stroke engines.[25]

Code Number	Description
1	Mixed Landuse patterns
2	Absence of NMV infrastructure
2a	Bicycle lanes, parking and loans available
3	Govt. run large bus fleet, deregulated in early 90's
3a	Skeletal bus service present
4	No restriction on two wheeler and car ownership
4a	Restriction on two wheelers
5a	Restriction on paratransit modes(cycle rickshaw, three wheelers)

Table 10.3. Description of City Environment

Delhi is showing declining trends of three-wheeler population because of a restriction on fuel and age of public transport vehicles. All three-wheelers are required by the orders of the court to run on CNG and should be less than eight years old. Shanghai and Beijing are the only cities in this list where a planned bicycle infrastructure exists. This may explain the high proportion of bicycles in these cities despite higher incomes than Indian cities.

Dependence on walking and cycling journeys, despite the absence of an infrastructure for them shows the presence of captive users of these modes. In the absence of organised public transport systems dependence on paratransit (cycle rickshaws and three-wheelers) increases. However, due to restrictive policies (legal regulations and restriction on movements) on these modes, the actual numbers of

[25] Zhou Hongchang, Daniel Sperling, Transportation in Developing Countries: Greenhouse Gas Scenarios for Shanghai, China, PEW Centre on Global Climate Change, July 2001.

these vehicles present on the road is often not included in the official registration numbers.

The entire transport system is not planned for in cities where the informal sector is an integral part of the urban fabric (Chinese cities were an exception until the 1980s). The two major trends that can be noticed are: (1) the predominance of walking, cycles and public transport, (2) cities which lack formal public transport services, are dominated by indigenously designed vehicles: three-wheelers in Indian cities, jeepneys in Manila, tuk tuks in Bangkok, cycle rickshaws in Dhaka and Kanpur. These are manufactured locally, and require minimal training to operate. They are another innovation for survival in the city, which is rewarding at an individual level, however, suboptimal at society level due to a lack of safety and environment concerns. It is also interesting to notice that, when restrictions are imposed on private vehicles like motorcycles in Chinese cities, their numbers remain low. However, restrictions on paratransit modes in Delhi, Jakarta and Dhaka have resulted in incorrect official statistics and individual strategies to circumvent the law. Passenger becaks (cycle rickshaws) are converted to goods becaks in Jakarta at the time of inspection. Pedestrians and bicyclists continue to use the roads despite a hostile environment because they are the captive users of these modes (except in Chinese cities).

Innovations for survival

Urban residents have shown a range of innovations to meet basic mobility needs. Often this has meant violating the formal plans and vehicle safety standards.
Delhi: Squatter settlements have grown close to the place of work inside the city. Most work trips are walking or by bicycles. Pedestrians and cyclists are present on all roads. Road hierarchy as per textbook definitions does not exist. People from resettled housing away from city centres travel in goods vehicles, violating safety norms. The actual number of cycle rickshaws in the city is estimated to be three times the current official figures.
Jakarta: Large number of *becaks* (cycle rickshaws) continue to exist in the city despite an official ban. Passenger rickshaws are converted

to goods rickshaws at the time of official checking. Each vehicle is shared by more than one driver to provide employment to more people. Three wheeled taxis imported from India in 1975 continue to run on the roads. Hawkers are found on the limited access expressway passing through the city.
Three-wheeled vehicles: *Autos* in Delhi, *bajaj* in Indonesia, *tuk-tuks* in Bangkok, *baby taxi* in Dhaka and larger vehicles, like *jeepneys* in Manila, *Tempos* in Indian cities, *Bemo* in Surabaya are locally manufactured. These vehicles do not meet safety or environmental standards, yet meet the mobility needs of the urban residents which have not been met by the formal public transport system.
Streets are used as per the local needs. Urban streets of Asian cities are used by pedestrians, non-motorized vehicles and motorized vehicles at the same time. However, the physical design is primarily influenced by the needs of motorized vehicles. In high-income countries it has been possible to some extent to define the primary functions of city streets and a well-defined hierarchy of streets along with suitable geometric designs have been developed accordingly. Often the urban arterial has a well-defined role of carrying relatively fast moving car traffic. The road cross section is designed to accommodate car traffic, and parallel service lanes are provided to meet the needs of abutting land use like shops or the entrance and exit from residential or commercial areas. Well defined shopping streets are often developed with large pedestrian paths, street shopping and roadside restaurants. In contrast to this, streets of Asian cities are often self-organized, multi-functional entities. The socio-economic conditions of most cities influence the way streets are used. This is reflected not only in the mix of traffic that is present on the road, but also the other activities which are present along the road. Arterial streets are used by motorised vehicles as well bicycles, cycle rickshaws and pedestrians. Street vendors, bicycle repair shops etc. are common sights on arterial streets and near bus stops. Often, streets in Asian cities are not restricted only to the movement of vehicles and people. The range of activities include services required by the users, and social activities which require safe public spaces.

Current policies

An urban transport trend that emerges as a result of streets dominated by pedestrians, non-motorised vehicles and paratransit vehicles meeting the accessibility and mobility needs of heterogeneous urban economy manifests itself in congestion, pollution and an increase in the numbers of accidents (Figure 10.4). The feedback loops suggests here that corrective actions should lead to changes in transport and activity system. Let us look at few successful and failed actions to gain insights for the future.

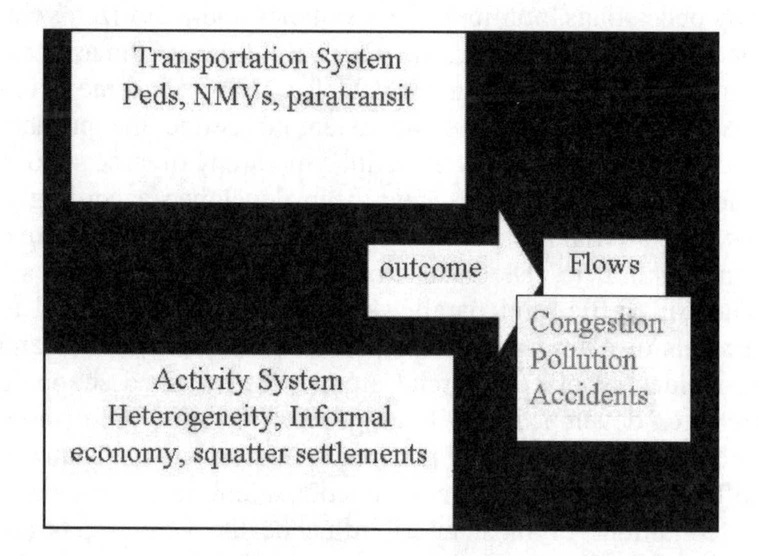

Figure 10.4. Relationship between Transportation and Activity System

Urban transport congestion solutions: The universally adopted solution in almost all megacities in Asia to solve the problem of congestion has been an investment in flyovers (grade separated intersections) to improve the speeds of motorised vehicles (cars and scooters). Pedestrians, cyclists and public transport buses do not benefit from these investments. In recent years Latin American cities have dealt with congestion in a different way. Space on urban streets has been allocated to public transport vehicles, creating high capacity bus systems.

Do we have any solution to reward the innovative skills of people who are forced to use the hostile infrastructure? Our understanding of use of urban streets and services required by street users is limited by the definition of legal, motorised modes. Yet, since the captive users have no choice, there is a mis-match between design and usage of infrastructure.

Solutions to urban transport pollution: the public transport fleet in Delhi has been converted to run on compressed natural gas. This has resulted in lower emissions of particulate matter. However, a large number of small buses have joined the fleet and the number of government run buses has declined. A formal public transport system has been replaced by an informal system where safety concerns have been compromised. The bus and three wheeled taxi fleet has reduced by 50%. The use of cars and two wheelers continues to grow. Why is it that pollution reduction strategies do not discuss the needs of zero pollution vehicles - bicycles and rickshaws? Investments in NMV (=non motorised vehicles) infrastructure not only reduces pollution but also improves safety, because the largest proportion of road traffic fatalities involve people not in cars in this region. However, policy-makers have not given any importance to such strategies. In the name of cleaning the environment, Delhi, Dhaka, Jakarta have imposed restrictions on rickshaws.

The success story of environment clean up has been the phasing out of lead. However, these technological fixes do not answer the mobility needs of captive users, the people without choice for whom restrictive laws have no meaning.

Future directions

A future solution lies in resolving the conflicts that exist at several levels. 1. Planning for the informal sector in the city. 2. Meeting the needs of captive users - pedestrians and cyclists prior to cars. 3. Ensuring safe accessibility and mobility to ensure a clean environment. This can happen when the problems of flow pass through the filter which is sensitive to these needs (Figure 10. 5). Methodologies and engineering solutions can evolve once a new paradigm is accepted.

An understanding of the symbiotic relationship existing between informal and formal sector is imperative for solutions. A poor understanding of the symbiotic relationship that exists between the formal and informal sectors leads to physical segregation, planned townships and slum relocations away from the formal city. Poor families who are also transport poor are denied accessibility to economic opportunities and exposed to the burden of reduced mobility.

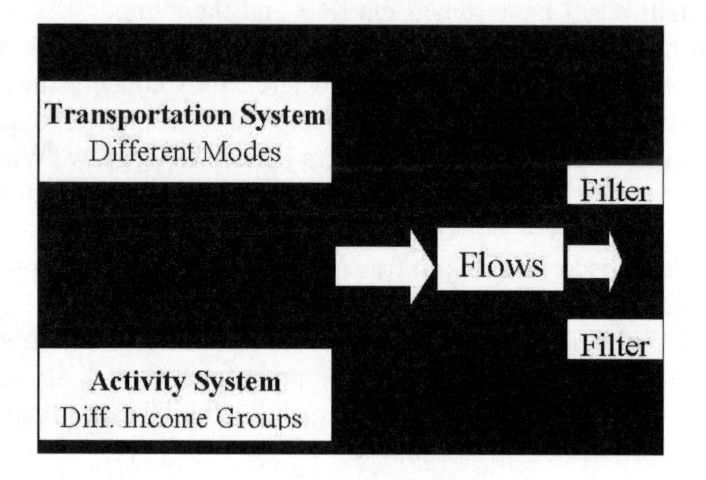

Figure 10. 5. Transportation and activity system interaction

Our current understanding of transportation issues leads to an 'improvement' in the transport situation by prioritising the movement of vehicles at the cost of lost public space. Improvements in road capacity in Asian cities has meant reducing pedestrian and bicycle facilities, removing street vendors, restricting pedestrian movements and constructing grade-separated junctions. This leads to lonely streets and spaces, which are of no interest to anyone, at the expense of substantial financial resources. Evidence from several Chinese and Indian cities shows that benefits to vehicular traffic is also short lived. On the other hand, if the informal sector is considered an integral sector of the urban economy, then an improvement in transportation involves changing the design criteria for urban arterials. The number and type of hawkers around a bus stop can be predicted on the basis of the number of waiting commuters. Cycle repair shops, cold drinks and snacks provided by street hawkers serve the same function as that of a well-designed service area along a highway. The difference is in the speed of users and, therefore, the difference in frequency and density of service providers. Thus standards for geometric design of these roads would include spaces for hawkers and the presence of hawkers will not be termed as 'encroachments'.

Lessons from self-organising systems for resolving conflicts

- The perception of benefits and risks can be influenced by adopting the right methodology (priority to pedestrians and NMVs)
- Better understanding of the function of streets, space utilization, etc. will give different design criteria
- The perception of mobility benefits and risks can be influenced by realizing that:
    - Access to work is the prime concern, not safe mobility
    - The benefits of increased speeds accrue to one sub-group, while the penalties are imposed on another

Self-organizing patterns of Asian cities provide us with insights for creating inclusive urban spaces that welcome diversity and meet the contrasting needs of different social groups. Future sustainability of cities and urban transport must learn from these regardless of the geographical location. This is central to the goal of building "A city for ALL."

Chapter 11. Urban Transport in Africa: Opportunities for Maximising Intervention in Complex Systems

Vinand M. Nantulya, Meleckidzedeck Khayesi and Wilson Odero

Introduction

The movement of people and goods in urban areas is essentially a time-space event that is associated with spatial interaction within and between areas. They arise due to location and separation of activities in time and space, and are conditioned by complementarities, intervening opportunities and transferability, as Ullman (1957) has clearly demonstrated in his model. The movements are expressions of the social organisation of space and of rational attempts to achieve effective integration between specific locations of human activity, including land use patterns, residential patterns, population densities, street geometry, location of work place, shopping precincts, health centres, and other traffic generation-attraction localities (Hilling 1996, Hoyle and Knowles 1992, Whitelegg 1987, White and Senior 1983, Lowe and Moryadas 1975).

Urban transport systems in Africa are complex. The inadequate infrastructure, multiplicity of transport modes that are poorly integrated and synchronized, the increasingly dysfunctional public transport systems and the resulting emergence of largely unregulated informal transport service, the multiplicity of agents involved in service provision (government, multinational corporations, and private sector), overall inadequate service provision for the different road users, weakly articulated and passive urban transport policy, and poor planning regulation and management contribute to this complexity. The deteriorating traffic congestion, increasing air pollution, and growing number of road traffic crashes and injuries are but a consequence of this underlying complexity and competition for space.

The search for solutions to the complex urban transport problems in Africa must take these issues into consideration, and there are unfolding opportunities in doing so. While acknowledging that urban transport in Africa is indeed complex, this chapter notes that there are certain unfolding opportunities both in the political arena, in the rapidly growing informal transport sector, and in south-south learning for addressing urban transport needs in Africa. Our study therefore examines ways that the opportunities for intervention can be exploited. But there is no quick-fix solution. It goes beyond just arguing that police should be effective in enforcing the law; or simply calling for replacement of small informal transport vehicles with buses and underground trains to reduce congestion, or the eviction of commercial bicycle operators from streets. The challenge is how to maximise the unfolding opportunities. Continuous engagement of all stakeholders is the key for success.

Conceptual framework

The urban transport system itself is complex with respect to development and organisation of different modes: walking, bicycle, minibus, bus and train. The different modes have different dynamics driving them. There is also a wide range of other factors to consider: infrastructure, flow systems, the amount of traffic generated, the trip making characteristics, trip origin-destinations, the service itself and management-planning for the transport system, and the tension between various providers as they compete for markets. Typically the "journey to work" is influenced by the location of employment opportunities, of industry, of residential area, and of schools; and the fact that we have a heavy flow of people in the morning and evening in one main direction. This requires the development of a transport infrastructure and service that is adequate in terms of the time schedules, the location of activities (serving various zones) and the heavy demand placed on the system at certain hours (the morning and evening peaks). In short, we are dealing with several supply and demand needs that have various points of origin and destination.

As Figure 11.1 shows, the transport system is embedded in a super-arching set of other relationships - the social, political, economic, and physical environment - which add to the complexity. Politics shapes, and can be shaped by, urban transport systems. Equally important, demand created by economic opportunity through industry, commerce and trade can shape the development and nature of urban transport services. The social sub-system is a consideration of such factors as population growth, demographic structure, settlement pattern and location of social services (recreational facilities, schools, hospitals etc) and the role they play in the provision of transport services. One of the current concerns is to provide suitable transport services for older persons, children and the disabled in urban areas in Africa. Physical environment - quality, ecology, terrain, landscape - all matter. The construction of transport routes also takes up space and affects the environment. Tensions within and among the over-arching socio-economic-political arenas influences the urban transport development and operational dynamics.

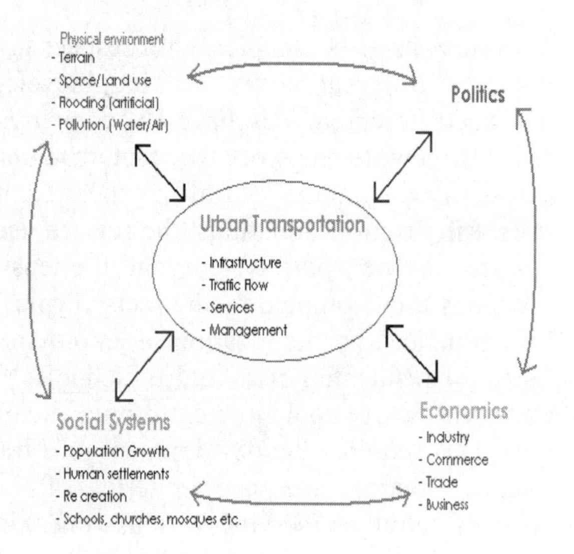

Figure 11.1. An urban transport system embedded in a super-arching set of other relationships

Opportunities for Interventions

Africa has historically been portrayed as a continent in a crisis, with little or no solution to offer for addressing its own problems and for the rest of humanity. An example of this portrayal is in a prescription for Africa's development dilemma that is depicted as a move or transition "from crisis to sustainable growth..."(World Bank 1989). This negative image makes it difficult for a number of people and institutions to realise that there are unfolding opportunities in the midst of what looks like a hopeless situation. We argue that the unfolding opportunities and positive changes are important, and that we should exploit them to develop interventions for addressing the complex urban transport systems in Africa.

Opportunities in the political arena

Africa has had challenges and instability in its political history since the colonial period (Gordon 1996). The settlement pattern of the colonial city and the attendant policies of the state led to the emergence of a patchy transport network and unequal service development in a number of African cities. The colonial transport routes were meant to effectively facilitate the transfer of resources from the colonial areas to Europe and also to get in resources to enhance and entrench colonial interests (Soja 1979). The development of the colonial urban centres on which the road and railway network was focused largely served the colonial interests of appropriation of resources from African countries (Kabwegyere 1979, Soja 1979). The colonial transport network that was founded to promote the attainment of the economic and political goals of the colonial state did not really serve local transport needs. The outward orientation of the African economies that started with the incorporation of African states into the global capitalist system contributed to the distortion of the African economy and corresponding infrastructure such as transport. This outward orientation has continued up to the present, perpetuated by strong political and economic forces such as unfavourable terms of trade for African goods (Rodney 1972, Hancock 1992, Ogutu 1993, Raheem 1996). The independence period has not substantially changed this structure.

There are, however, some positive changes taking place in the African political arena. There are ongoing reforms and/or discussion of reforms. The one-party state system is gradually giving way to multi-party politics. There has been reference to a second liberation and an African renaissance. There are deliberations on urban governance and the need to reform this area. The point we are stressing is that the reform taking place in Africa provides an opportunity to advance the cause of urban transport. For example, in Kenya, there are efforts to set a new political agenda since the new government came to power in December 2002. There are constitutional talks and debates going on. The challenge to urban transport experts is to use this opportunity to push transport agenda into this process. In Nairobi there is a new mayor who has shown interest in reforming the urban system, dealing with crime, dealing with accessibility, reorganising the informal transport sector that caters for the bulk of public transport needs. Where are the local and international players to support this effort?

Opportunities are unfolding in Kampala, Uganda. The city council is trying to generate revenue and this is partly done through commissioning bicycle and taxi operators' associations to do this (Howe and Davis 2002, Kinyanjui and Khayesi 2003). What kind of revenue generation strategy that involves the transport sector can we put in place to help the local authority? It is important to get the city council to consider not just collecting revenue but implementing other positive mechanisms to tackle the transport crisis. Where are we in this movement? Soon after President Museveni of Uganda came to power in 1986 he mobilized commuters in privately operated taxis to organize themselves into stakeholder groups so as to regulate, in a demand-driven manner, passenger overloading. The result was an outstanding success. When he later issued a directive that any driver running over a pedestrian at a pedestrian crossing must be taken directly to his office for discipline, there emerged a culture that respects pedestrian crossings throughout the city of Kampala. Can this be applied elsewhere?

In Ghana the government initiated its first urban transport project in 1993 (Kwakye et al. 1997). This programme has involved dialogue with the public. There is much more that can be tapped from such a venture with respect to research and interventions.

In South Africa the apartheid dimension of urban transport and transport in general is fairly well documented (Khosa 1989, 1993, 1995). Since the democratic elections of 1994 the government has taken steps to redress the immense apartheid iniquities, one of them being transport. The government position and effort is spelled out in the "White paper on transport" (The National Department of Transport, Pretoria, Republic of South Africa 1996). This was followed by production of a 20-year strategic framework for the transport sector. This framework is known as "Moving South Africa – The Action Agenda" (The National Department of Transport, Republic of South Africa, 1999). There is a recent paper that outlines achievements in the transport sector within the framework of sustainable development (The National Department of Transport, Republic of South Africa 2002). These are blue prints requiring great effort in order to translate them into practical solutions. What are we doing as urban transport researchers and practitioners to support this initiative? Are we reviewing and giving feedback to this document? Are we funding any of the interventions proposed?

In short, the political reforms taking place in different African countries offer opportunities for interventions in urban transport. The reforms may be comprehensive and/or specific. The challenge before us is: What are we doing as urban transport experts to enlarge on this reform process to make sure that it takes on board the transport complexities that we are talking about? Is there a possibility of taking the high level conferences organised by the Volvo Educational and Research Foundations a step further by convening country level or regional stakeholder consultative forums that enrich and support the emerging opportunities in different regions of the world such as Africa? Is there a possibility for the Volvo Educational and Research Foundations supported Centres of Excellence to play critical facilitative and catalytic roles in different regions of the world, including Africa? Do we have such a centre in Africa? What form should it take?

Thus, the political process itself is unveiling certain opportunities. Urban transport planners and scholars should scan the political landscape much more critically and find ways of operating in the political arena, and of working together with the political managers in assessing the evidence and developing viable solutions to the complex urban transport problems in Africa. As transport scholars and practitioners, we need to start analysing the agendas of political parties and relevant governance issues that have a bearing on urban transport to identify opportunities for intervention. Transport is a political investment. Political reasons are an important consideration in transport infrastructure and service development (Ogonda 1986). To what extent have we pursued the political dimension in our analysis and interventions? How do we engage the political process? We have good findings, but how do we situate them in the political process and change that is taking place now?

Opportunities presented by the growing informal transport sector

Informal transport services are a key feature of the transport sector in developing countries (Cervero 2000). In Africa these transport services have partly emerged in urban and rural areas to fill up a transport vacuum emanating from neglect and collapse of the government-provided transport service (Rizzo 2002, Khayesi 2001, Howe and Bryceson 2000). These services are diverse: small public transport vehicles, bicycles, motorcycles, hand-carts and animal-drawn carts. This sector is crucial in facilitating the movement of people and goods as well as offering employment to a number of people in Africa (Kinyanjui and Khayesi 2002, Rizzo 2002, Howe and Davis 2002, Khayesi 2001). The people employed are predominantly youths from a low socio-economic bracket (Nantulya and Musiime 1999). This sector cannot be neglected in dialogues and planning for the complexity of the urban transport in Africa. There are a number of problems associated with this sector, especially the mini-bus: violence, road traffic injuries, strong political-economic vested interests, pollution and social influence. But this is only the down-side of a system that caters for under-developed urban and rural areas, and offers transport for poor populations (Nantulya and Reich 2002).

There are interesting features in this sector: organisation and management, relationship with the state, local entreneurship and stakeholding. There are some African cities that depend to a great extent on the informal transport services. For example, in Dar es Salaam, the mini-bus (commonly referred to as *Dala Dala*) has a 98% share of total public transport trips (World Bank 2002). In Nairobi, the *matatu* has a 70% share of public transport trips and in Addis Ababa, the mini-bus has a 72% share of total public transport trips (World Bank 2002).

In Uganda the taxi system has grown in importance for both intra- and inter-urban transport. The urban bus service has been out of existence for a fairly long time in Uganda. The mini-bus taxis, largely based on local capital, have filled this vacuum. A recent attempt by one wealthy local entrepreneur with the support of an external investor to introduce buses for urban transport was spectacularly unsuccessful. The taxis out-competed the buses. In Kenya the bus system in Nairobi and Mombasa is facing stiff competition from the mini-bus (locally known as *matatu*). A government-operated bus service introduced during the colonial period and offering transport service in urban areas as a monopoly, has collapsed partly due to this new competition, although mismanagement also played a role. It is instructive that a large bus company in Nairobi is making an effort to introduce small buses to operate like *matatus* so as to withstand the competition. But these are receiving stiff resistance from *matatus*.

In a number of African countries, such as South Africa, Ghana, Ethiopia, Botswana, Ghana and Nigeria, the informal small-scale public transport vehicle service is important in urban transport. Also growing in importance is the use of commercial bicycle transport in African cities. A good example is in Ugandan towns, where there is a mix of bicycle and motorcycle transport services (Howe and Davis 2002).

In addition to the modes described above, walking, hand-drawn carts and animal-drawn carts feature in African transport services. These modes fulfil short-distance trip purposes and are critical to certain segments of the urban population. Thus, Africa is not only a walking rural continent but also a walking urban continent.

The scenario presented above reveals the complexity of issues that have to be considered when planning for urban transport in Africa. The main problem at the planning level is that, generally speaking, the urban transport planners in Africa do not take into account the dynamic transport scenario that exists. They are still fixated on planning for the motor vehicle. Though commercial bicycle and motorcycle transport is continuing to grow in importance in Kampala, Jinja and other major towns in Uganda, there has been very little change in road transport planning to provide lanes, parking facilities and credit for this mode of transport that is estimated to be partly sustaining the livelihood of about 1,600,000 people (or 7% the population). Whitelegg (1993) argues that Africa has a unique opportunity to promote sustainable transport, especially in the planning and promotion of green modes of transport: walking and bicycle.

However, the growth of the informal transport sector has not been easy sailing. In East Africa it grew against state monopolies and police harassment. Until recently there have been frequent strikes by *matatu* operators along a number of routes in Kenya, protesting against police harassment and demands for bribes (Khayesi 2002). There are also unconfirmed reports that the corruption practised by the junior traffic police officers has the support of some of the senior ranking officers in the same department, as they share in the take at the end of the day. While we do not condone this act at all, we have been reflecting on it and asking ourselves: does this not reflect a much more complex issue than the mere physical exchange of money between the policeman and the motorist? Our reflection has led us to probe the following issues: What is the social and economic reality under which an urban traffic policeman lives? What are the terms and conditions of work in the police force? How much does a policeman earn? Who bothers about the welfare of this police officer? How good is the licensing system for vehicles and drivers? We have gathered anecdotal evidence that cause us to argue that a corrupt policeman is part of the complexity that surrounds the urban transport system and governance in Africa. He is a victim of this complexity, and he also contributes to the complexity.

Are there opportunities for intervention in this sector? Yes there are. But a re-orientation in perception and approach in urban transport planning is needed. These services need to be perceived as part of the process of urban transport development rather than as an unwelcome step-child of mainstream urban development. Are African urban transport planners and leaders equipped with the relevant expertise to plan for these kinds of services? We have described how bus transport systems have collapsed due to the stiff competition from the informal sector. But the informal sector is quite organized in its own ways. So could the operators of privately owned vehicles used in the informal sector be facilitated to start to invest together and develop a more coordinated urban transport system in a bottom-up approach? Would such a broad ownership platform stand a better chance of successfully introducing the use of buses in the urban transport system, as opposed to a one-entrepreneur attempt in Kampala?

Opportunities from south-south shared learning

There has been a general problem of urban transport knowledge and technology transfer. In fact, we have had more of knowledge and technology transplant than genuine and meaningful adaptation to the local circumstances. There are many urban transport consultancy reports which are not based on a strong working partnership between foreign expatriates and local experts and institutions. We, therefore, need to rethink this process and explore opportunities in south-south exchange and influence. We shall use two examples: the Bogotá model and the Indian Institute of Technology.

Bogotá is a city in Colombia. Previously, it was on a downward spiral in virtually all regards. But since 1998 a change started to take place (see Box 11.1). We have quoted the example from Bogotá at length to show that change is possible and can be realised. This model can be promoted in cities in Africa. It is an example of both a promising intervention in urban transport, and also of the role of urban governance in facilitating change. Documenting and monitoring the experiences and trends in the Bogota model, disseminating the findings, and organising targeted and interactive policy and governance in selected African cities to initiate and support any promising directions towards improvement of urban transport can have a positive impact. Who will do this? Could the Volvo Research and Education Foundation facilitate this process?

"In Bogotá, Colombia, Enrique Penalosa, the mayor from 1998-2001, held a referendum and reallocated transport budgets to improve the quality of life for the very poorest. The results were staggering. The city embarked on an intensive programme of building cycling and pedestrian-only routes, including a car-free route, 17km long, connecting some of the poorest parts of the city with the facilities they need to access, including jobs. Parks were built on derelict land, canals cleaned up and car-free days implemented. In October 2000 the citizens of Bogotá voted in favour of excluding cars from the city in the morning and afternoon peaks from 2005. Penalosa introduced a car numberplate system that required 40% of the cars to be off the roads during peak hours on two days a week, and this produced a reduction in pollution. More than 80 miles of main roads are now closed for seven hours every Sunday and, each week up to 2million people come out to enjoy the clean air, the freedom and the safe environment. In 2002, a car-free day was set up on a weekday and 7million people went to work without a car. In a subsequent poll 82% supported the concept. Bogotá's approach is based on creating an equal and vibrant city where no one need fear the oppression that pervades so many other developing countries' transport systems. Penalosa wanted a reliable and free-moving bus system that was affordable and used road space on the surface. An underground or metro, he reasoned, was simply too expensive for a poor country and, in any case, was supported only by rich people because it keeps intact as much road space as possible. Now the buses carry more than half a million people every day, are reliable and affordable, and give the poorest groups in Bogota as much accessibility to jobs and facilities as the rich. The bus system also covers its costs and makes a profit, while every metro in the world swallows up huge subsidies, which are further losses from health, education and sanitation programmes."

Box 11.1 The Bogotá model. Source: Whitelegg (2003)

As Whitelegg argues, "traditional transport policies simply do not work for the poor - whether in Colombia or Britain. Western countries can learn from experiences such as this and we should stop sending our transport consultants to developing countries. We need the radical approach pioneered in Colombia, with its emphasis on equality, democracy, openness and citizen participation - especially of women, the elderly, children and those who walk, cycle and travel on buses" (Whitelegg 2003: 24). The Bogotá model provides an example of potential south-south and south-north exchange and influence.

Another example of south-south shared learning is the Indian Institute of Technology (IIT) based in Delhi, India (http://www.iit). There are a number of relevant research projects and programmes undertaken at this institute in the Transportation Research and Injury Prevention Programme (TRIPP). There are at least two fundamental lessons from this programme that are relevant to Africa. The first is the approach utilized. TRIPP has adopted an interdisciplinary and integrative approach whereby there is an effort to examine issues related to transportation in their entirety. IIT-TRIPP has brought together researchers from different departments at the Indian Institute of Technology to work on specific themes. Furthermore, IIT-TRIPP works closely with government departments, industry and international agencies. The lesson to learn and promote from this approach is the contextualisation of urban transport in Africa, as we have attempted to do in the conceptual-analytical model presented in Figure 11. 1 (see also chapter 10).

The second lesson from TRIPP is the relevance of research and intervention projects. We shall cite one example. TRIPP is playing a critical role in conducting studies and planning for a high capacity bus system in Delhi. This project is focused not just on the bus but on the entire mobility environment: walking, cycling, parking, sheds, street vendors and land use planning. This project will be launched soon, making New Delhi one of the few cities in the developing world to implement a high capacity bus system. The main lesson to draw out for African cities is the need to plan for all road users and attend to multiple needs such as providing space for traders. TRIPP can interact and enrich the expertise in urban transport planning and project implementation in African countries.

Conclusion

Urban transport in Africa is complex, but there are unfolding opportunities - in the political arena, in the growing informal transport sector and in networking, and learning from other countries and experiences in and out of Africa. To understand the breadth of this challenge, key characteristics of the urban transport systems in Africa are noted: inadequate infrastructure, multiplicity and mix of transport modes, poorly integrated and synchronized public transport service, increasingly multiplicity of service providers, unorthodox competition for space and market, weakly articulated urban transport policy and poor planning and regulation. These characteristics form the background for opportunities in the political arena and opportunities in the informal transport sector. The challenge is therefore to maximize existing opportunities, engage stakeholders and secure networking in Africa (south-south, north-south). Examples of opportunities in the political arena are: ongoing political and economic reforms, urban governance – a new generation of mayors and political leaders are getting involved directly, for example in Uganda, South Africa and Ghana. Examples of opportunities in the informal transport sector, as seen in Africa as a whole are: in Addis Ababa the mini-bus has a share of 72% of the total public transport trips; in Uganda, it is the main system for transporting members of the public; in Dar es Salaam, the mini-bus (*dala dala*) has a share of 98% of the total public transport trips, and in Nairobi, the *matatu* has a share of 70% of the public transport trips.

The conclusion of this chapter is that while urban transport in Africa is indeed complex, there are certain unfolding opportunities in the political arena, in the rapidly growing informal transport sector, and in south-south learning for addressing urban transport needs in Africa. The challenge is how to maximise on these opportunities. Continuous engagement of all stakeholders is the key to success.

References

Cervero, R. 2000. *Informal Transport in the Developing World*, Nairobi: United Nations Centre for Human Settlements (Habitat).

Gordon, D.L. 1996. "African Politics". In Gordon A. A. and Gordon, D. L. (eds) *Understanding Contemporary Africa*, London: Boulder. Chapter 4 (pp. 53-90).

Hancock, G. 1996. *Lords of Poverty*, London: Macmillan.

Hilling, D. 1996. *Transport and developing countries*. London: Routledge.

Howe, J. and Bryceson, D. 2000. *Poverty and Urban Transport in East Africa: Review of Research and Dutch Donor Experience*, International Institute for Infrastructural, Hydraulic and Environmental Engineering.

Howe, J. and Davis, A. 2002. "Boda Boda – Uganda's Rural and Urban Low-capacity Transport Services" in Xavier Godard and Innocent Fatonzoun (eds.), *Urban Mobility for All,* Lisse: A. A. Balkema Publishers, pp.235-240.

Hoyle, B. S. and R. D. Knowles, eds. 1992. *Modern Transport Geography*, London: Belhaven Press.

Kabwegyere, T. B. 1979. "Small Urban Centres and the Growth of Underdevelopment in Rural Kenya", in *Africa*, (49) 3, pp.308-315.

Khayesi, M. 2001: "Matatu Workers in Nairobi, Thika and Ruiru: Career Patterns and Conditions of Work", in P. Alila and P. O. Pedersen, (eds.) *Negotiating Social Space: East African Micro Enterprises*, Trenton: Africa World Press, Chapter 4, pp. 69-96.

Khayesi, M. 2002. "Struggle for socio-economic niche and control in the matatu industry in Kenya", in *Development Policy and Management.* Vol. IX, No. 2.

Khosa, M. M. 1989. "Dipalangwang: Black commuting in the apartheid city, Republic of South Africa", in *African Urban Quarterly,* Vol. nos. 3 & 4, (August/ November), pp. 322-329.

Khosa, M. M. 1993. "Transport and the 'Taxi Mafia' in South Africa", *The Urban Age* 2(1), pp. 8-9.

Khosa, M. M. 1995. "Transport and popular struggles in South Africa", in *Antipode: A Radical Journal of Geography* 27, no. 2, pp. 167-188.

Kinyanjui and Khayesi. 2002. Social Capital in Micro and Small Enterprises in East Africa: Implications for Poverty-Alleviation. Final Research Report Submitted to the Organization for Social Science Research in Eastern and Southern Africa (OSSREA), Senior Scholars Research Program, Addis Ababa.

Kwakye, E. A. et al. 1997. "Developing Strategies to Meet the Transport Needs of the Urban Poor in Ghana", in *World Transport Policy and Practice*, Volume 3, Number 1, 1997, pp. 8-14.

Lowe, J. C. and Moryadas S. 1975. *The geography of movement*, Houston: Houghton Mifflin Company.

Nantulya, Vinand, M. and Florence, Muli-Musiime. 1999. The road traffic accidents in Kenya: Health equity and the policy dilemma. Kchapter.

Nantulya, V. M. and Reich, M. 2002. "The neglected epidemic: Road traffic injuries in Developing Countries", in *British Medical Journal*. 324, pp. 1139-1141.

Ogonda, R. Timothy. 1986. *The development of road system in Kenya*, Ph.D thesis, University of Nairobi.

Ogutu, Z. A. 1993. "Sustainable Development in Sub-Saharan Africa: What are the Alternatives" in *Journal of Eastern African Research and Development*, Vol.23, pp. 24-39.

Raheem, T. A. 1996. "Western NGOs in Africa: Bodyguards of the Advancing Recolonisation", in *NGO Monitor*, pp.13-15.

Rizzo, M. 2002. "Being Taken for a Ride: Privatisation of the Dar es Salaam Transport System 1983 – 1998", in *Journal of Modern African Studies*, 40, 1, pp. 133-157.

Rodney, W. 1972. *How Europe Underdeveloped Africa*, Dar-es-Salaam, Tanzania Publishing House.

Soja, E. 1979. "The Geography of Modernisation-A Radical Reappraisal", in Obudho R.A. and Taylor, D.R.F., eds., *The Spatial Structure of Development: A Study of Kenya*, Colorado: Westview Press, pp.28-45.

The National Department of Transport, Pretoria, Republic of South Africa .1996. White Paper on National Transport Policy (available at (http://www.polity.org.za/govdocs/white_papers/transwhite.html).

The National Department of Transport, Pretoria, Republic of South Africa .1999. Moving South Africa – The Action Agenda (http://www.transport.gov.za/proje…-agenda-may99/actionagenda01.html)

The National Department of Transport, Pretoria, Republic of South Africa .2002. Sustainable Transport for Sustainable Development (www.transport.gov.za)

Ullman E. 1957. *American commodity flow*, University of Washington Press.

White, H. P. and Senior M. L.1983. *Transport geography*, London: Longman.

Whitelegg, J. 1987. "A geography of road traffic accidents", in *Trans. Inst. Br. Geogr.* N.S.12, pp. 161-176.

Whitelegg, J. 1993. *Transport for Sustainable Future: The Case for Europe*, London: Belhaven Press.

Whitelegg, J. 2003. "Bogota takes fast lane to fairer transport", in *The Guardian.* 12-18 June 2003.

World Bank. 1989. *Sub-Saharan Africa: From crisis to Sustainable growth*, World Bank, Washington.

World Bank. 2002. Scoping Study: Urban Mobility in Three Cities – Addis Ababa, Dar es Salaam, Nairobi, SSATP Working Paper No. 70.

Chapter 12. Examples of Positive Changes of Complex Urban Transport Problems in Europe

John Whitelegg

Introduction

One of the many remarkable aspects of European urban transport is the enormous gulf between the best examples and the worst and the enormous gulf between the high quality of many offerings and the deeply ingrained negative attitudes of many citizens. European urban transport is an excellent example of a special kind of "gulf war syndrome". We are not very good at addressing the issues raised by these "gulfs" but we are getting better and the central role of sustainable transport in achieving a wide range of economic, social, environmental and quality of life objectives is now widely appreciated.

The first gulf (best versus worst) is important because the existence of poor quality anywhere encourages negative attitudes and encourages the political support of car-dependence and car use as the only realistic transport policy option. In Manchester (UK) the suburban rail system provides some of the worst public transport experiences anywhere in Europe. Most railway stations on the main line to Manchester Airport (eg Mauldeth Road, East Didsbury) are in very poor condition, with no information on train running and no staff. Trains are cancelled and passengers on the station are not informed. The stations are lonely and perceived of as dangerous. The tickets on the trains are not valid on buses and vice versa. There is no concept of integration with bus services. Over 100,000 residents of south Manchester experience very poor quality of public transport provision supported by totally inadequate bicycle and walking infrastructure. This is the reality that permeates many minds, including the minds of journalists and politicians when they think of public transport.

Only 5% of trips to Manchester Airport (20 million passengers per annum) are by train because the trains are perceived of as so unreliable, expensive and disconnected from any concept of well staffed, safe and integrated operation.

Fortunately there are many examples of well run, integrated, well staffed and reliable services across Europe. Travelling by metro in Madrid or Vienna is a pleasure. Travelling by bicycle in Copenhagen and most Dutch cities is safe, enjoyable, cheap and health promoting. Walking in many European cities is still of poor quality because of the priority given to vehicles and the automatic downgrading of pedestrian priority. Even in Copenhagen 6 lanes of traffic swirl past the Town Hall on Hans Christian Andersen Boulevard.

The fact remains that we have a large number of very successful urban transport projects and administration in Europe all of which provide best practice examples of how to improve the economy, the environment, quality of life, sustainability and health. The challenge for all of us is to ensure that this is the norm in Europe and this will involve (for many) some very uncomfortable thinking about fiscal rebalancing and highway space reallocation. I return to these later.

The other gulf (citizen perception) is now much more clearly understood than before. In the UK we have 14 "individualised marketing" demonstration projects which engage directly with citizens with high quality information and discussion and incentives. This idea is based on the work of Werner Broeg in Munich and on his Australian projects (especially Perth). In York (UK) I am responsible for one of these projects and we have made direct contact with 5,600 citizens to discuss how they can make more use of buses, walking and cycling. The results of all 14 projects will be available in April 2004. The most important result so far is that citizens are willing to engage in this process and are willing to change behaviour. This underscores other work carried out in the UK on travel plans. In our travel plan work (eg with Pfizer, the US pharmaceutical company and with a large hospital in Plymouth) we can "shift" 15% of the car user away from the car. There is a willingness to change travel behaviour.

In this paper I want to emphasise the successes of European urban transport policy and highlight the areas that need more work in order to bring about a mobility transition. At the core of this transition is a shift away from the car and all its impacts on physical form and liveability towards mobility based on the widest possible considera-tion of all the alternatives to the car and on the overriding impor-tance of civilisation, liveability and sustainability.

Introductory Examples of European Best Practice

There is already a considerable body of experience on achieving significant modal shifts and the associated traffic reduction in Euro-pean cities and regions:

- Lemgo in Germany has increased bus usage from 40,000 to over one million in one year
- Zurich in Switzerland has held levels of auto ownership and traf-fic volumes constant for a decade whilst public transit use has soared. Limited investment in road capacity coupled with serious priority of public transport vehicles, reduced parking place num-bers and the "Zurimobil" concept
- Houten in the Netherlands has developed a comprehensive bicy-cle-pedestrian network and cut car trips per household by 25 per cent
- Swiss and German research on car-sharing shows that people who have joined a car sharing scheme (not car-pooling) and who have previously owned a car have reduced their car mileage by 50 per cent. The Federal Ministry of Transport in Germany estimates that car sharing will reduce annual vehicle kilometres by 7000 million. In Europe as a whole the figure is put at 30,000 million vehicle kilometres reduction
- In Aachen (Germany) traffic into the city centre has been reduced by 85 per cent over the last ten years, the car's share of transport has gone down from 44 per cent to 36 per cent and NOx pollution has gone down by 50 per cent

- In Bologna a deliberate policy of traffic restraint involving the closing of streets and park and ride produced a 48 per cent drop in motorised traffic entering the historic core and a 64 per cent drop in cars (1982-1989)
- In Groningen (Netherlands) in 1990 48 per cent of all trips within the city were by bicycle, 17 per cent on foot, 5 per cent by public transport and 30 per cent by car
- In Manchester the Metrolink tram has taken up to 50 per cent of car journeys off roads in the area it serves. It has replaced over one million car journeys into the city centre each year
- Five per cent of car users switched to a new “City Express” bus service in Belfast in the first 6 months of operation
- Edinburgh has set itself a traffic reduction target of 30 per cent
- In Leicester 10 per cent more 7-9 year olds were allowed to walk to school after traffic calming
- Levels of cycling in one of the “Safe Routes to School” pilot projects have more than doubled even without the necessary infrastructure works being carried out. More than 120 pupils at Horndean Community School in Hampshire are regularly cycling to school compared with about 50 last autumn and just 36 when the project began in at the end of 1995
- The “Carte Orange” in Paris covering all modes and introduced in 1975 led to a 36 per cent increase in bus patronage. The London travel card led to a 16 per cent increase in public transport use at a time of decline elsewhere

Selected Case Study Material

Planning Process

John Roberts (of the consultancy TEST) ran a comparison of Almere (The Netherlands) and Milton Keynes and demonstrated the extent to which land use and transport planning can influence the demand for motorised transport: “the most obvious finding and an important one, was the much higher percentage of trips made by car and the much lower level of bicycle use in Milton Keynes when compared to Almere (65.7 per cent of trips by car compared to 43.1 per cent, 5.8 per cent of trips by bicycle compared to 27.5 per cent respectively)”.

The influence of compact cities on reducing motorised trips is reviewed in Smith, Whitelegg and Williams (1997). Physical land use planning is a tried and tested method of reducing the length of trips, increasing the use of non-motorised modes and reducing the demand for expensive road infrastructure.

Parking

Restrictions in mainland European cities such as Zurich and decisions as in Amsterdam to reduce car park numbers (Lemmers, 1996) provide best practice examples. Good practice parking policies exist in Sheffield, Winchester, Leeds, Southampton, Cambridge and Edinburgh. An MVA consultancy study of Bristol for the Department of the Environment, Transport and Regions shows that car trips into central Bristol can be cut by 41 per cent by a 75 per cent reduction in on-street parking, higher charges and enforcement of planning permission for non-residential parking.

Reallocating Space and Modal Preferences

There are many isolated examples of successful policies in this area: the Manchester Metrolink, bus lanes in several British cities, Zurich's prioritisation of public transport; Maidstone Integrated Sustainable Transport (MIST) project; car-free residential and city centre areas (Lubeck, Amsterdam, Berlin, Edinburgh); building homes on car parks; bicycle priority schemes and planning in York and Cambridge, Delft and Groningen (the Netherlands), Detmold and Rosenheim (Germany); Copenhagen's cycling strategy; Darmstadt's (Germany) encouragement of cyclists and pedestrians to share the same large car-free space in the city centre; SMART buses in Liverpool; new tram systems in Strasbourg; innovative car-sharing initiatives (StattAuto) in Berlin, Bremen and Edinburgh (e.g. 3000 participants in the Berlin car sharing scheme have removed 2000 cars from the roads of Berlin). Vienna has adopted a policy of constructing several hundred extended pavements at crossings and tram stops to improve safety for pedestrians.

Traffic Management

Groningen (Netherlands) has developed a sector access model; Bochum (Germany), has prioritised its trams in preference to cars; Gothenburg (Sweden) has divided the central business district into 5 cells which has had the effect of reducing car mobility by 50 per cent; Houten (Netherlands; pop. 30,000), has given preference to bicycles, restricted access by sectors and traffic restraint. Over the last 20 years Oxford has produced one of the lowest rates of traffic growth in the city centre of any UK city, through parking controls and Park and Ride schemes.

Marketing

Large scale marketing exercises have increased bus patronage in Lemgo (Germany). Similar but less ambitious schemes can be found in the UK, for example SMART buses (Liverpool) and TravelWise schemes. System-wide, discounted tickets have helped increase public patronage in Germany, for example the *Umweltkarte* or "environment tickets", as in Freiburg where the *Umweltkarte* is attributed with a reduction of 4000 cars per day on the roads to the city centre.

Green Commuter Strategies

These are increasingly common in the UK, for example in Nottingham (City, County, Queens Medical Centre, Universities and Boots), Plymouth (Derriford Hospital), and Oxford (University planning agreement). The Rijnstate Hospital in the Netherlands has restricted its car parking provision to 400 spaces for 2050 staff. Transport Demand Management policies have increased the use of public transport from 8 per cent to 40 per cent of all journeys. Restricting car parking availability was the key to this success. The Pfizer travel plan (Sandwich, UK) has reduced car use by 15% and included "parking cash out" rewarding every member of staff with 3 Euros per day when they arrive at work and deducting the same amount at the end of the day if they leave by car with one occupant only (bus users, pedestrians, cyclists and car sharers keep the cash incentive and do so every day).

Identification of Policies with Traffic Reduction Potential based on fiscal intervention

Traditionally transport policies have been based on strong fiscal intervention in support of car ownership and use. In the UK the real cost of car use has declined at a time when public transport costs have increased dramatically and valuable car parking places in cities are still provided to many groups of employees free of charge (eg public officials in Swindon, UK park for free in the city centre when one space has a market value of 1200 Euros per annum). This fiscal bias can be reversed/managed to bring about a more sustainable transport future for urban areas. Examples include:

- Urban road pricing (including congestion pricing)
- Fuel taxation
- Taxation on parking spaces at the workplace
- Taxation of parking spaces at car-intensive developments e.g. out-of-town and edge of town retailing complexes, airports, leisure centres, sports facilities
- Regional norms on car parking provision to deter a competitive bidding-up process
- Financial support from hypothecated revenues for quality public transport, co-ordination, integration, dense pedestrian and cycle networks and innovative programmes of accessibility enhancement for rural areas
- Substantial policy integration at the national level so that transport and land use planning policies support health policies, climate change policies and vice versa
- Modification of the "predict and provide" approach which still determines policies towards airport capacity and housing provision
- Legislation that will provide for the establishment of regional transport authorities following the German models. These authorities will be charged with the responsibilities of bringing public and private finance to bear on the supply of public transport, high quality integration and co-ordination , "environment ticketing" schemes and high quality information
- Providing new methods of funding public transport e.g. fuel taxation as in Germany and employer contribution as in Paris

- Eliminating subsidies to private motorised transport through the company car, business mileage and corporation tax regimes

Not all these policy areas will be discussed here.

Road Pricing and Fuel Taxation

Road pricing and fuel taxation is an attractive policy option, particularly if the revenues could be recycled into the local economy to support all the alternatives to the private car. According to the OECD (1995) survey of transport policy options road pricing is being considered in some shape or form in most OECD countries. Plans are well advanced in Cambridge and Edinburgh (UK), toll systems exist in Norway, Stockholm is planning to introduce such a system and road pricing was introduced in London in February this year. Road pricing is generally suggested for those locations where the growth rate in traffic is already the lowest across a number of geographical situations. The growth of traffic into and out of central London has been far lower than the growth in outer London or the growth on the M25 corridor. Road pricing is best seen as a strongly supportive measure alongside a battery of other measures including strong land use controls and modal preference.

The London congestion charge is still the subject of intense discussion and monitoring but it is already clear that it has reduced the number of vehicles crossing the cordon line by up to 30% each day. This has improved bus running times and made conditions for pedestrians and cyclists much easier. Paradoxically the success of the measure has led to a reduced income stream and funding problems for public transport and safety measures. Transport for London has indicated that underground fares will increase by about 25% in the near future which will make these fares the highest in Europe. The mayor of London also intends to build a new 6 lane road bridge in east London (the Thames Gateway Bridge) with associated access roads which will generate thousands of extra vehicles each day.
The view of the OECD (1995:154) is that "The key to the sustainable development strand is a substantial and steadily increasing fuel tax coupled with (other) measures". The UK has had a policy com-

mitment to increase fuel tax by 6 per cent above the rate of inflation at each annual budget. This has now been abandoned. The OECD suggest that the impact of a 7 per cent p.a. rise in fuel costs in real terms would be to "quadruple fuel prices in 20 years … [leading[... to lower car ownership levels compared with what they would otherwise be, fewer car trips and shorter trip lengths". An overall reduction in car trip-making of about 15 per cent, a reduction in trip length of about 25 per cent and an overall reduction of vehicle kilometres of one-third is predicted if fuel prices rise by a factor of 2.5 (OECD, 1995: 156).

Traffic Reduction for Lorries

HGVs are a longstanding problem in towns and cities, on trunk roads through villages and in or near national parks. In general their impact is much greater than their numbers would suggest. Their impact on noise, road damage, pedestrian and cyclist fears and air quality is large and there is a strong case for reduction in ways that can protect the economy of towns and cities and the consumer who has come to depend on goods and services supplied by HGVs. Considerable progress has been made in this area in mainland Europe, particularly Germany, whilst hardly any progress at all has been made in the UK. In Germany HGV reduction strategies which pay attention to the commercial interests of the companies involved are generally referred to as "City-Logistik" strategies.

City Logistics involves setting up new partnerships and styles of co-operation between all those involved in the logistics chain and in delivering/receiving goods in city centres. These partnerships offer significant reductions in vehicle kilometres and truck numbers and are currently in existence in Germany and Switzerland. City Logistics are a very clear illustration of the importance of developing high quality organisational arrangements and inter-company co-operation agreements in addition to whatever new technology might be appropriate. City logistics have taken transport operations into an area of development that builds links and emphasises co-operation across all players and interest groups. In Germany partnerships between logistics contractors are reducing lorry numbers and improving the urban environment. These partnerships (known as City Logistik companies in Germany) are in operation in Berlin, Bremen, Ulm, Kassel and Freiburg. The Freiburg example has several pointers to the future shape of freight transport in urban areas.There are currently 12 partners in the scheme. Three of the partners leave city centre deliveries at the premises of a fourth. The latter then delivers all the goods involved in the city centre area. A second group of five partners delivers all its goods to one depot located near the city centre. An independent contractor (City Logistik) delivers them to city centre customers. A third group, this time with only two service providers specialises in refrigerated fresh products. These partners form an unbroken relay chain, one partner collecting the goods from the other for delivery to the city centre.

The Freiburg scheme has reduced total journey times from 566 hours to 168 hours (per month), the monthly number of truck operations from 440 to 295 (a 33 per cent reduction) and the time spent by lorries in the city from 612 hours to 317 hours (per month). The number of customers supplied or shipments made has remained the same. The Kassel scheme showed a reduction of vehicle kilometres travelled of 70 per cent and the number of delivering trucks by 11 per cent. This has reduced the costs of all the companies involved and increased the amount of work that can be done by each vehicle/driver combination.

These reductions in vehicle numbers and in traffic levels have benefited the companies through higher levels of utilisation of the vehicle stock. It is not in the interests of logistic companies to have expensive vehicles clogged up in city centres, one-way systems and on circuitous ring roads. There are clear economic benefits arising from lorry traffic reductions.

Conclusions

There is no shortage of really good ideas and really good practice in European urban transport. There are still a lot of traffic jams, there are still a lot of lorries carrying milk from Denmark to Germany and

Germany to Denmark and there is still a lot of fiscal bias in favour of the car and the lorry. More importantly there is still a lot of political confusion and bias. The following is an extract from the official position of the Conservative Party in the UK on transport issues:

> [the labour government] has made the car the most expensive part of the family budget. Labour has ignored the fact that poor access to quality modes of transport cuts off parents, patients and all passengers from the economy, and indeed, society more generally. This Government has already tried and failed to price drivers off the road. What it has failed to realise is that people will still use their cars if public transport cannot take the strain. The next Conservative Government will put an end to this political war against drivers.
>
> The Labour Government's record has been truly appalling on road building, having provided the smallest number of road schemes in any period since World War Two. In 2001 not one inch of tarmac increased England's road network, for the first time since tarmac was invented in the 1860s. Yet UK taxpayers contributed to improving roads in Scotland, Italy and elsewhere in the EU - even in Iraq.

An e-mail from the Leaders Office, UK Conservative Party, London, 14th August 2003 runs:

> There is still a huge gulf between the reality of what has been achieved and what can be achieved in European cities and the mindset of many politicians. It would be foolish to think this problem has gone away and it would be even more foolish to think that it won't get worse.

Section four: The Role of Actors in Coping with Urban Transport Development

Introduction

The fourth part of this anthology focuses on the role of various actors attempting to solve the problems of urban transport systems are. Among these actors are consultants, political decisions-makers, policy-makers, planners, administrators, industrialists, financial agencies, representatives of NGOs and, sometimes, researchers (when they recommend a certain policy). The role of the actors is, of course, crucial but at the same time their actions are hampered by existing institutions, by current power relations, and by values dominating among the general public.

In the first chapter the question of various barriers is addressed from the perspective of a consultant. The author of the chapter, Desmon A. Brown, has made a study of what he calls "the 'non-technical' issues that create barriers to successful implementation of urban transport projects in developing countries". His empirical basis is taken from the city of Kingston in Jamaica but he believes that his results are also representative of other cities in developing countries.

A questionnaire was completed by people who work (or have worked) in the Land Transport section of the Transport Ministry of Jamaica. They were asked to rank a number of significant reasons why effective transport solutions have still not been implemented over the past 20-40 years. Some key individuals were subsequently interviewed. The results of these two inquiries were finally reviewed by the author.

Desmon Brown identified four groups of barriers: 1) political 2) policy related 3) economic and 4) social barriers. Some examples follow. *Political barriers*: Brown often found a lack of political will to take the necessary actions required for successful implementation.

The nature of many political systems encourages the implementation of projects that produce results mainly in the short term. *Policy related barriers*: There is an absence of clear transport policies and specific transport development plans. The role of different modes of transport and their importance for the mobility of the citizens is not well understood. Sometimes there are conflicts between transport policy and the policies of other government agencies.

Economic barriers: The link between transportation and economic opportunities is often badly defined. This is particularly serious as transportation is a substantial employer of semi- and unskilled labour. *Social barriers:* The social stigma of public transportation is often underestimated. The link between transportation and accessibility to social and other opportunities is badly defined.

Finally Desmon Brown claims that the results of his inquiry are relevant for political institutions, policy formulators, implementers and individuals who may be affected by the measures recommended. He also takes the view that international funding agencies should demand greater accountability from transport consultants for their recommendations.

In the second chapter of this section, Ethan Seltzer and Andy Cotugno present an inspiring example of successful planning from Portland in Oregon, USA. Their basic view is that "transport system development outside of a broader context provided by goals for sustainability and community quality of life is ultimately self-defeating". The title of their chapter is therefore "Making the land use, transportation, air quality connection". Their somewhat rhetorical presentation is based on the experiences of the regional government, Metro. One of the stated aims of Metro is to develop "a truly multimodal regional transportation system".

The authors emphasize the role of planning in the regional development process, probably a necessary statement in an American context. But they also point to "the legitimacy of employing collective action to address the failure of the market and existing institutions". They are, therefore, eager to include widespread participation in local planning efforts, claiming that such participation (from the part of ordinary citizens, house builders, business interests and environmentalists) increase the range of ideas in development.

The role of strong leadership in the development process is regarded as a key factor by the two authors. This role is, however, clearly changing over time. The authors believe that "the leaders of the future will be those able to create the partnerships and collaborations needed to advance overall community values". In the Portland case, the endeavour of future leaders is currently guided by a concrete vision called the "Region 2040 Growth Concept". The creation of this vision was a result of citizen participation in the planning process, coupled with the work of planners and elected officials.

The next chapter in this section has been written by Stephanos Anastasiadis. He brings an environmentalist's perspective to the discussion, having been closely associated for a period of time with the environmental organisation, "Transport and Environment" ("T&E"), a Brussels-based umbrella organisation of national environmentalist organisations all over Europe.

At the beginning of his contribution he describes his vision of a future compact European city characterised by a sustainable and environmentally clean transport system (where the Polluter Pays Principle is finally accepted). Accessibility is a more important value in such a city than a high level of mobility. Socially disadvantaged citizens are not excluded from this accessibility. A city of this kind would, according to Anastasiadis, become a cultural and economic magnet. The way to such a future has to be constructed within a political framework consisting of the city level, the regional level, the national level and the European Union level.

He then enumerates the ten main barriers to creating sustainable transport systems already identified by the European Conference of Ministers of Transport (ECMT). He elaborates on some of these;
1) political unwillingness to address the problems of urban transport development, 2) the existing complex organisational structure (the

four levels mentioned above), 3) existing mobility patterns created by habits and historical circumstances rather than by rational choices, 4) people's perceptions of different modes of transport and – as always , 5) vested interests in the present transport system.

In his own analysis of possible ways to realize a sustainable and culturally and economically vibrant European city, he focuses on the potential of the EU level. He finds, however, that, within the Union there is a paralysing conflict not only between the two Directorates of the European Commission responsible for transport and environment policies respectively, but also between the principle of Subsidiarity and the Commission's fulfilling its Treaty obligations on promoting health and the environment. This conflict can only be overcome through strong political will - both at the EU-level and at the level of the member-states. "The biggest challenge of all is in breaking patterns of beliefs and patterns of behaviour."

In the final chapter of this section, Carmen Hass-Klau and Graham Crampton discuss ways of changing urban transport systems in a more limited sense. However, their case is an example of how researchers may also be seen as actors (not only as analysts) when recommending particular transport policies.

Their study is an analysis of the economic impact of investment in light rail. The introduction of light rail is today often recommended for environmental reasons. The two authors show, however, that the economic effects of new lines seem to be underestimated by many decision-makers.

They have based their study on analyses of investment in light rail in 15 urban areas, most of them in France, Germany and the UK, but also in North America (three cases). They are able to show that some direct effects of rail investment have do to with the values of properties and rents near to light rail stations. Other, more indirect, effects are associated with retailing turnover in the city centre and with the profitable location of business.

Investment in light rail may also contribute to a reduction of car use and car ownership (for instance by eliminating the need for a second car in middle class households) as well as a reduction of car parking spaces in the neighbourhood of offices and business centres. In both cases, it is a question of a substantial saving to households and business.

Hass-Klaus and Graham Crampton claim that close co-operation between private developers and planners is absolutely necessary for success in choosing the new alignment of a light rail line. Co-operation of this kind, based on better knowledge of the potential economic effects of the investment, is an attractive option for the future of urban transport development. Finally, the two authors list a number of key lessons to be learnt from their study.

Chapter 13. The "Non-Technical" Issues that Create Barriers to Successful Implementation of Urban Transport Projects in Developing Countries

Desmon A. Brown

Introduction

Many developing countries have been the subject of countless transport studies, reviews, updates of previous studies and project proposals. In most cases these are sponsored by international funding agencies, such as the World Bank or the Inter-American Development Bank, usually as a prerequisite for receiving funding for transport-related projects.

A detailed review of these studies/proposals will reveal that the recommendations of many of the recently commissioned studies are only updates of previous studies/proposals, confirming that the original recommendations were for the most part technically sound. Nevertheless, the recommendations of a substantial number of these studies were never fully implemented. Why then has effective implementation eluded so many transport projects?

One obvious result of the non-implementation of sound transport plans in urban areas is the increased congestion observed in many cities. The increasing access to private motor vehicles over the past two decades has led to increasing traffic congestion, not only in developed countries but also in most developing countries. With so many studies and so much information currently available on transport research, why have the results of these studies not addressed the ever-increasing congestion problems?

The problems appear to be more acute in developing countries as not only are the existing infrastructure woefully inadequate, but there is also a lack of sufficient funding to implement the changes required. Is this perception correct? Were these issues not addressed by the many studies or are there other factors, "non-technical issues", which have had a significant effect on the implementation of these studies/proposal.

In seeking answers to the above questions, Jamaica will be used as a case study. A review of what has transpired in Jamaica over the past forty years should help in identifying these "non-technical issues" that have created barriers to successful implementation of urban transport projects. It is proposed that these "non-technical issues" are representative of other developing countries.

This paper presents the results of a study that will attempt to examine the causes of non-implementation, despite the technical veracity of the recommendations of these studies/ proposals

Background

Jamaica is the largest English-speaking island in the Caribbean area and its capital is Kingston located on the south coast.

Over the past 10 years, Jamaica has been "flooded" with used vehicles, mainly from Japan, as that country was mandated to replace its vehicle fleet with more ozone-friendly units. The availability of these relatively cheap units has resulted in a large increase in vehicles operating in Jamaica.

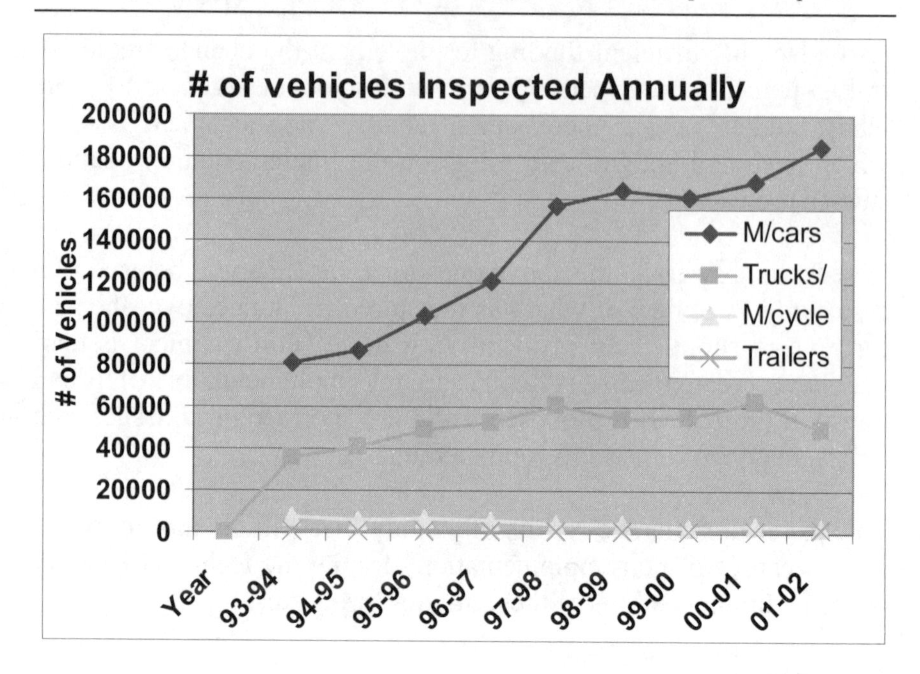

Graph 13.1. Number of vehicles inspected annually in Jamaica

Graph 13.1 shows that the number of motor cars examined by the state vehicle examiners doubled between 1993 and 2001 with an overall 90% increase in motorised units operating on the roads. During this same period there was no significant expansion in the road infrastructure and the resulting traffic congestion is evident not only in the capital but in all the major towns.

Methodology

Some research was conducted to ascertain answers to the questions posed. The methodology employed was as follows:

- Conduct a survey among individuals who are still involved or who used to be involved in transportation policy planning or implementation to obtain empirical information
- List the transport-related studies/proposals in Jamaica within the last 20-40 years

- Examine these studies/proposals to identify possible reasons why effective transport solutions have still not been implemented
- Make a review based on observation and discussions with persons presently in or who were previously connected to the transportation industry

The Survey

A questionnaire was completed by a number of individuals who are presently in or had previously worked in the Land Transport section of the Transport Ministry or one of a number of related departments and agencies.

Respondents were asked to rank the following reasons in order of priority from 1 to 20. The most significant reason should be given priority 1, the next 2, etc. They were instructed that if they disagree with any of the reasons listed, they should *not* rank that reason by placing a number next to it. In addition, five blank spaces were included for the respondents to insert additional reasons not listed.

The respondents were also asked to list the number of transport-related studies that they were aware of.

Srl.	REASON	Ranking 1-20
1	Recommendations technically unsound	
2	Recommendations not applicable to Jamaica/KMTR	
3	Recommendations did not include an implementation plan	
4	Government bureaucracy	
5	No funding provided for implementation	
6	The cost to Jamaica to implement recommendations was considered too high	
7	Recommendations not in line with existing Government Policy at the time of the recommendations	
8	Political interference	
9	Inadequate regulation of the transport industry	
10	Inability or unwillingness of the implementers or the Government to make the difficult changes called for by these recommendations	
11	Person tasked to implement policy was ineffective	
12	Person tasked to implement policy showed little interest in implementing the recommendations.	
13	Other social reasons	
14	Other economic reasons	
15	Security reasons	
OTHER REASONS (List)		
16		
17		
18		
19		

Table 13.1 Ranking of reasons of barriers to successful implementation of urban transport projects

The Review

Discussions were held with a number of key individuals, some of whom are no longer employed at the Transport Ministry. Obtaining statistical information on why such plans/ recommendations were unsuccessfully implemented was extremely difficult and as such this aspect of the review was somewhat limited. Information was difficult to obtain for the following reasons:

- The majority of the individuals with detailed knowledge of these studies were some of the very persons responsible of the implementation
- Most of the individuals still employed at the implementing agencies declined to participate in the interview
- Different persons were involved in the implementation process over the period in question and had varied influence on these projects

Results of survey

At least forty-nine (49) transport-related studies were identified by respondents to the survey as being carried out over the last 40 years. Specific findings were:

- 54% stated that they knew of 4 or less studies
- 38% stated that they knew of between 5 and 8 studies
- 8% stated that they knew of more than 8 studies
- 77% stated that they had an input in some of the studies
- 92% said that they were satisfied with the recommendations of the studies that they were aware of
- 70% of the respondents said that were satisfied to some extent, whilst 30% said they were not satisfied to any extent, with the implementation of the recommendations of the studies

The following graphs show the main reasons given by the respondents for non-implementation or ineffective implementation of recommendations:

- Reason 10 (Inability or unwillingness of the implementers or the Government to make the difficult changes called for by these recommendations) at 46%
- Reason 6 (The cost to Jamaica to implement recommendations was considered too high) at 15%

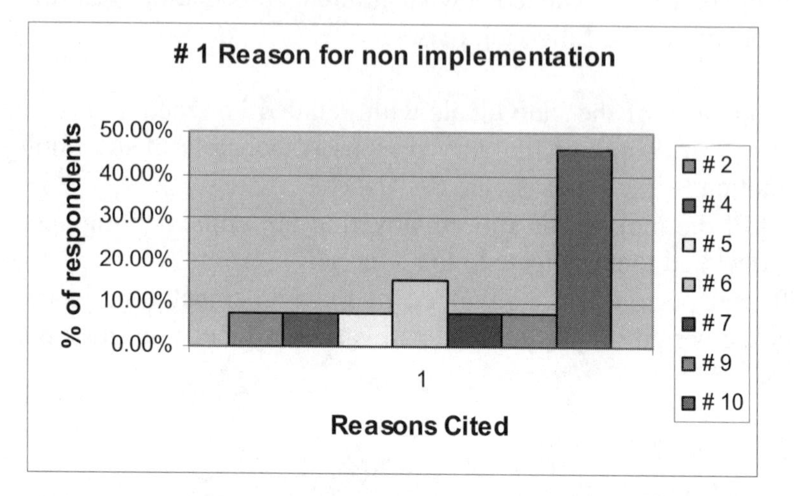

Graph 13.2. The primary reasons of barriers to successful implementation of urban transport projects citied by the respondents

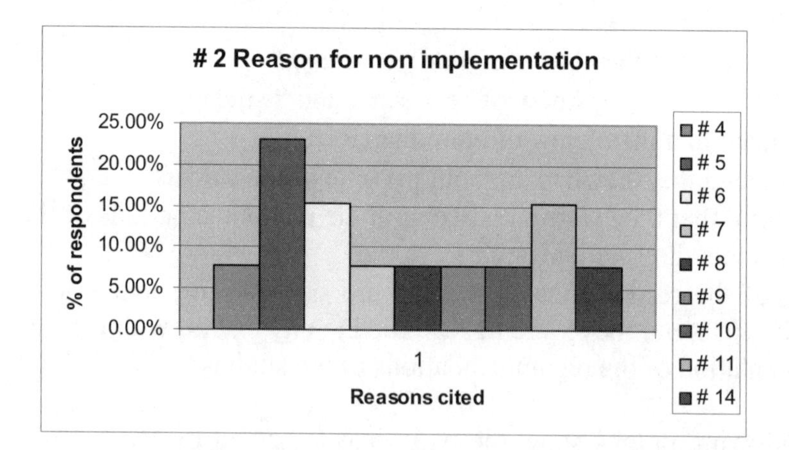

Graph 13.3. The second ranked reasons citied by the respondents

- Reason 5 (No funding provided for implementation) at 23%

- Reasons 6 & 11 (The cost to Jamaica to implementing recommendations was considered too high and the person tasked to implement policy was ineffective) at 15%

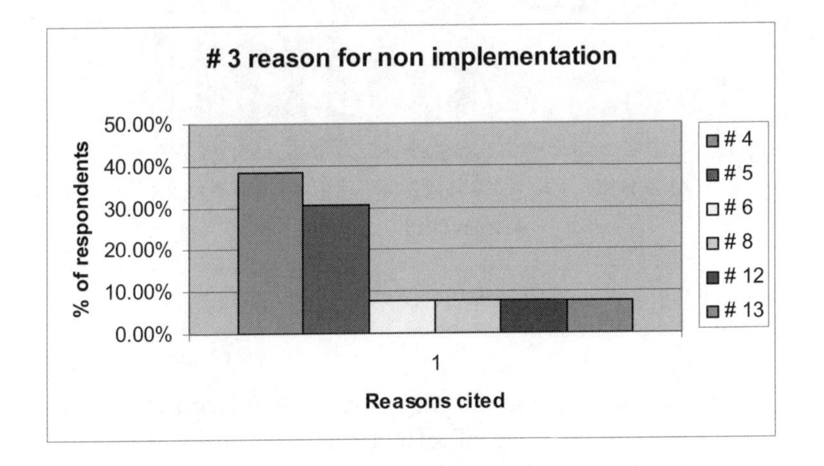

Graph 13.4 The third ranked reasons citied by the respondents

- Reason 4 (Government bureaucracy) at 38%
- Reason 5 (No funding provided for implementation) at 31%

To determine the overall significance of each reason, 20 points were allocated for each time a reason was cited as the 1 reason, 18 points for the 2 reason, 16 points for the 3 reason etc, etc and 2 points for the 10 reason. No points were allocated for reasons that were not rated between 1 and 10. Graph 13.5 shows the overall significance of each reason cited.

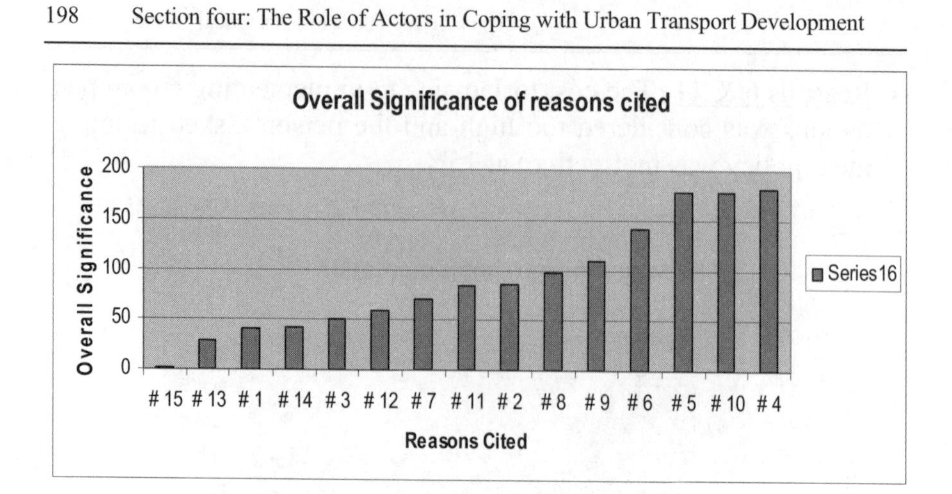

Graph 13.5. Overall significance of reasons citied

- Reason 4 (Government bureaucracy) was considered by respondents to have the greatest overall effect
- Reasons 5 & 10 (No funding provided for implementation and inability or unwillingness of the implementers or the government to make the difficult changes called for by these recommendations)

A few respondents added a number of issues which in their opinion significantly affected the effective implementation of transport projects (Graph 13.6)

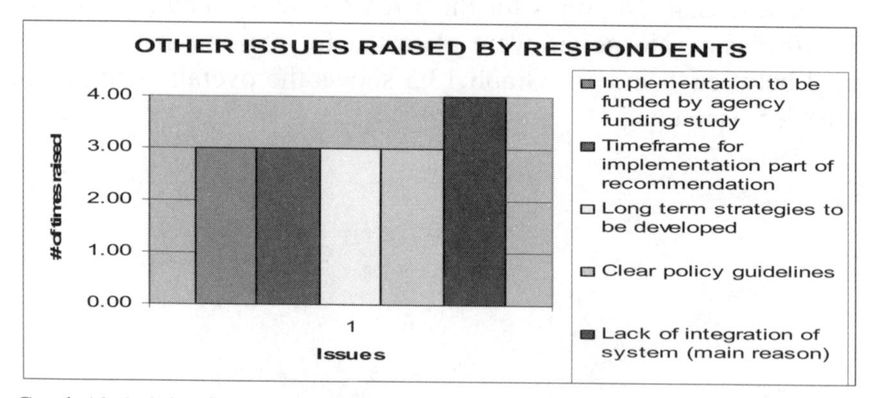

Graph 13.6. Other issues raised by respondents

Review and personal observations

The non-implementation of the recommendations of the many studies/proposals must be seen within the context of the economic impact/implication on society. The provision of transport services, especially as it relates to employment opportunities, must not be overlooked. This fact takes on even greater importance in developing countries where the economy comprises a significant informal sector. Globalisation, coupled with inappropriate economic policies, has negatively impacted on many developing countries, resulting in many lay-offs and redundancies. The provision of transport services provides an outlet for many of these displaced persons, especially since the business is conducted on a cash basis. Any dislocation of these individuals so "employed" would constitute a major political problem. In addition, the implementation of the infrastructure component of transport projects must also take into account the individuals (many occupying land illegally) who are normally displaced by these major projects.

Transport infrastructure projects offer numerous employment opportunities, especially for semi-skilled and unskilled persons. These projects are usually very attractive to the political directorate and are therefore highly ranked for implementation. As a result, the infrastructure segments of many projects are in many cases the only ones implemented.

Despite this, the many transport-related studies – in many cases the National Transportation Policy from which specific short-, medium- and long-term plans are developed – are irrelevant. Unfortunately, in a number of developing countries there are no specific transport plans and physical development is therefore a result of uncoordinated plans, implemented usually as a result of loans/grants from international funding agencies or government-to-government assistance programmes.

In quite a number of cases, those tasked with implementing road infrastructure work have had very negative experience of public transport (usually due to the failure of the local public transport system to deliver quality service). As a result, road improvements tend to give priority to maximising vehicle (motor car) traffic flow and overall improved mobility and accessibility is not the overriding factor. It is also observed that where there are no specific transportation plans in place, the incidences of "intervention" by the policy-makers in the implementation process are more evident.

The link between transport policy and other government policy (environment, energy, health, education, housing etc.) and their interrelatedness is usually not clearly defined and in some cases not understood. A number of examples exist where a country's effort to solve a major housing problem without adequate consideration of transportation issues (and other policy issues) have led to major transportation problems.

This "misunderstanding" of the inter-relationship of transport policies with other government policies has led to different agencies implementing plans that are in conflict with each other. An example of this is the recent conversion by the City Council of every unused government-owned lot in the New Kingston Central Business District into cheap parking lots (in an effort to increase revenue), which in fact encourages more people to use motor-cars for commuting. Arguably, a National Transport Policy would help to address this issue.

Another issue is that it appears that the link between accessibility by the poorest in society to social, recreational and employment opportunities on the one hand and transportation on the other has not been clearly defined.

The role of city governments in the development process and subsequent implementation of transport policies and plans has been negligible. An interesting observation has been that in many cities with effective and efficient transportation systems, in both developed and developing countries, a relatively strong city government is in place and it has been the city government that has spearheaded the implementation of the changes required.

The absence of an effective, tried and tested public transportation system in general has led to the shunning of public transportation by the affluent in some developed and nearly all developing countries. In these countries nearly all "professionals" commute by private motor car. This stigma will be difficult to overcome, especially in cities with additional problems such as a high crime rate and extreme weather conditions. Innovative solutions, such as park & ride systems utilising high quality air-conditioned buses, coupled with an extensive public education and public relations campaign along with the relevant technical solutions, must put across the recommendations.

The implementation of many transport-related projects without a clear understanding of what constitutes "success" is evident in many instances. A prime example of this was the privatisation of the public transport system in Jamaica in the 1980s. This process was lauded as a "success" by the international agency that recommended and supported this move, although the resulting public transport services provided by the private sector were nothing short of a disaster.

Conclusions

The non-technical barriers to successful implementation of the recommendations of the many transport-related studies conducted in Jamaica - and by extension in most developing countries over the last 30-40 years - can be categorised under the following four headings.

- Political
- Policy-related
- Economic
- Social

<u>Political</u>

- There is in many cases a lack of political will to take the necessary and in most cases difficult actions required for successful implementation of transport-related projects. This is usually due to the perceived negative political effects of such actions, especially when such actions are not effectively incorporated into an overall implementation strategy

- In most cases implementation is limited to segments that offer employment, especially to semi-skilled and unskilled workers
- The nature of many political systems encourages and gives priority to the implementation of projects that can produce positive results in the short term

Policy-related

- Despite the many studies there is an absence of a clear transportation policy and therefore specific transport development plans (short, medium and long term) which the Policy document should facilitate are also absent
- There is not always a clear understanding of the relationship between the need to provide adequate mobility for citizens, the role of different modes of transport in this process and their effects on congestion
- The non-coordination of transport policy with policies of other government agencies can result in outright conflicts in some cases
- There is in many cases a complete misunderstanding of some of the transport issues by policy-makers, resulting in frequent intervention in the implementation process

Economic

- Transportation and related industries are substantial employers of semi-skilled and unskilled individuals, especially in the informal sector. Any recommended changes that may affect employment opportunities must be carefully managed and addressed as an integral part of the recommended solutions
- The link between transportation and economic opportunities must be clearly defined for development purposes

Social

- The social stigma of public transportation usage is not to be underestimated and plans to bring about a reversal of these attitudes must be a central part of any implementation plan and must include a major public education programme

- The link between transportation and accessibility to social and recreational opportunities must be clearly defined for development purposes

What is abundantly clear is that transport consultants must pay far more attention to these "non-technical" issues if their recommendations are to be effectively implemented. On the other hand, international funding agencies must demand greater accountability from these consultants and ensure that ongoing transport studies do not become just pre-requisites for a new grant or loan.

Recommendations

1. Technical consultants must:

- Re-tool themselves to review these "non-technical" issues as part of their consultations
- Include solutions to these "non-technical" issues in their recommendations

The recommended solutions must address the following four target groups

Srl.	TARGET GROUP	ISSUES	POSSIBLE ACTIONS
1	Political directorate	- Identifying the main issues - Ensure understanding of the links with other policies - Sensitising to and full understanding of some of the difficult decisions that will have to be taken	- Briefing sessions to, among other things, identify the current views of the political directorate in respect of transportation issues and possibly follow up with a short paper with alternative views - Outlining the problems after identifying them, with possible alternative actions and allowing the political directorate to have an input early in the discussions.

2	Policy formulators	- Same as above for the political directorate, although more time should be spent addressing the following: - Links with other policies - Importance of planning for the short, medium and long term. - Benefits to the country of an effective transportation system	Same as above for the political directorate, although more time may have to be spent on this group. Possibly a symposium or a special training programme for relevant individuals could be arranged and examples from other countries used to strengthen proposals. A cost/benefit analysis of proposed improvements should be developed as part of the strategy to encourage change
3	Persons that may be affected by recommendations	- Those who may be affected negatively should be identified early on in the study and the economic impact quantified	- Where possible plans should try and incorporate these individuals. May require additional training for some - Where incorporation is not desirable, possible alternative employment should be investigated to minimise impact and potential opposition to implementation
4	Implementers	- This group should be included in as many of the above discussions as possible.	- This group should be identified as early as possible in the study and an implementation strategy developed jointly. The matter of incentives for switching to public transport should be discussed and the implementation team should include PE/PR professionals.

Table 13.2 Recommended solutions: target groups, issues, possible actions.

2. International Funding Agencies should;

- Assign equal importance to implementation as they do to "technical" solutions
- Ensure that implementation plans are a necessary part of any study
- Ensure that governments commit to the implementation plans *before* any funds are released
- Ensure that this commitment is either in the form of a Contract, a Heads of Agreement or a Memorandum of Understanding

3. Governments of developing countries must;

- Make the paradigm shift that prioritises the *full implementation* of projects
- Partner with international lending and donor agencies in "Capacity Building" to ensure that sufficient skills and technical knowledge are available to effectively manage the implementation of projects

4. Prior to the commencement of any transport project, the following should be clearly established and agreed on by all parties concerned;

- What are the factors that would constitute "success" in the specific project?
- What measurement tools would be utilised?
- How and when will these measurement tools be employed?

Chapter 14. Portland and Smart Growth: What you can Learn from the Portland Experience

Ethan Seltzer and Andy Cotugno

Introduction

As the nation turns its attention to growth management and smart growth, the magnifying glass has been focused on Oregon and the Portland-Vancouver metropolitan area. The Oregon statewide land use planning program is a 30-year work in progress, longer if you include earlier less successful precursors. It is known for the use of urban growth boundaries, preservation of farm and forest land, and the rigorous use of comprehensive planning shaped from outside the local community by statewide planning goals.

The Portland region is known for its regional government, Metro, its sustained effort to develop a truly multimodal regional transportation system, including the use of light rail, and its collaborative approach to regional growth management that exacts mutual accountability from all jurisdictions within Metro's jurisdiction, south of the Columbia River. Long a source of interest and sometimes inspiration for planners and regionalists, the Portland experience is now drawing the attention of commentators ranging from the Congress of New Urbanism to the Reason Public Policy Institute to the Natural Resources Defense Council.

In addition to the national attention, the history and current initiatives in Portland also draw attention from an international audience. The author has personally spoken to groups from the Netherlands, Budapest, Prague, several Japanese metropolitan areas, and Copenhagen within the last several months.

Sometimes, particularly in the days before the 2000 United States Presidential election, it seems like everyone has an attitude on smart growth and Portland's experience, or at least they're trying to. The conversation at the national level in 2000 became progressively more shrill, starting with Al Gore's embrace of smart growth on the campaign trail, and George Will's broadside aimed at both Gore and smart growth in the Wall Street Journal.

Today, the debate between those in support of smart growth and those in opposition still simmers. It is a familiar one for most planners: the legitimacy of employing collective action to address the failure of the market and existing institutions to produce liveable, resilient settlement patterns in our metropolitan regions. Regardless of how you feel about either the suburbs or the cities, grassroots citizen initiatives to enact growth boundaries, growth controls, and growth management efforts stand as testimony to the fact that as the nation's population becomes increasingly metropolitan, the nation's metropolitan areas aren't working very well.

To be sure, some of the forces at work are not all that laudable. Citizen desires to avoid contact with people different than they are, in terms of race, ethnicity, and socioeconomic standing, still exerts a strong push and pull on the desire to control growth. However, in addition to these longstanding dynamics in American society are relatively new concerns regarding environmental degradation, sense of place, community capacity, and urban design. Rather than the dawn of an era where place would become interchangeable, we are in fact beginning to understand that place has become both a matter of choice and a signature element in defining local, even household relationships to the emerging global knowledge economy.

As more communities seek new approaches to defining the challenges of late 20th-century urban development and addressing them, attention is turning energetically to smart growth and regional growth management. As this discussion leaves the dining room for the school auditorium, enters city hall and even stalks the halls of the Capitol in Washington, D.C., the search is on for "best practices", and in our experience, that search is turning up the heat here in Portland on what our experience has been and what it might mean for us and for others.

10 Lessons

If you are thinking of visiting us, or as you consume the analysis of pundits, keep in mind that every community must make its own successes. There is no software available for making one place just like another, because no place is really like any other. What we've done has emerged from who and what we are. The urban growth boundary works for us, but it is certainly not the only planning tool for limiting sprawl and preserving resource land.

If you visit or study us, look carefully not just at how we've done things, but why we've done them. That for us is where the true test lies, and for you, too. The point is not that you ought to employ the tools that we've used, but that you ought to select ones that best fit the problem you're trying to solve and the kind of place that you come from. Nonetheless, take heart. Much can be done, and our experience offers up ten lessons worth discussing in your home town:

<u>1) This is not an Experiment</u> - Often the Portland experience is described as an experiment. It's not. We are doing what we are doing because it's important to us. This region did not set out to become a national model. Instead, it sought to serve the values that have consistently characterized this community: a real desire to make up our own minds and solve our own problems, and significant concern for the environment.

This is an intentional place, and what we've done is the result of an ingrained desire to serve it well. As former Governor Tom McCall, "father" of the Oregon statewide land use planning program, once said, "Heroes are not giant statues framed against a red sky. They are people who say: This is my community and it's my responsibility to make it better." This is also an apt description of what has kept us going.

We are a product of our history, landscape, and current circumstances. This is one of the oldest continuously inhabited places in North America. For over 10,000 years our region has been a rewarding place for human settlement. Our interest in and involvement with national environmental and cultural movements has had a profound impact, as has our location just north of California, the fifth largest economy in the world.

Simply stated, time goes by and there are no controls in this business. You can't go back and "run the experiment" again with a different set of parameters. Consequently, this is not an experiment just as what is happening in your community is no experiment either. There is no such thing as doing nothing. Every move you make and don't make is part of what you will become. Places can choose a future, but they cannot choose not to choose. You might as well articulate your choice accurately, and actively make it your choice, since you'll be living with the consequences anyway.

<u>2) Planning Matters</u> - Through planning you can change the patterns that rule your landscape, and you can make things happen. As Lew Hopkins, Gerrit Knaap, and their colleagues have shown, plans matter. The experience that we've had in downtown Portland demonstrates that planning matters. In the early 1970's, a revolution of sorts was brewing in Portland. Urban renewal efforts of the late 1950's and 1960's had destroyed city neighbourhoods in the name of trying to save them. Planning in the mid-1960's called for even more highways, again as a means for saving neighbourhoods.

As one activist of the time has said, after protesting the Viet Nam war for the better part of a decade, they read Jane Jacobs' *The Death and Life of Great American Cities* and decided to save the city. A new generation of City Commissioners got elected at the end of the 1960's, among them a charismatic young lawyer by the name of Neil Goldschmidt, destined to become mayor a few years later. A proposal to build a 10-story parking structure at the 100% corner in downtown Portland galvanized a protest movement in opposition to the planning and plans of the time.

As a means to provide certainty, downtown landowners and business interests turned to planning and the incorporation of the rising tide of dissent into the direction for downtown. The result was the 1972 Downtown Plan, widely hailed as the blueprint for the ongoing turnaround for downtown Portland at the time that other downtowns were losing their grip. Some 25 years later, the success of the downtown plan was recognized through the development of a walking tour of its highlights, a celebratory dinner attended by over 700, and a day-long event to set the course for the next 25 years that drew the participation of 200 over the months leading to the event and 450 on the day that it took place.

Downtown Portland remains a work in progress, a celebrated work in progress. Planning crystallized the vision, made roles and responsibilities clear, and provided common ground that has sustained nearly three decades of investment and public policy. The old dictum "make what you want easy and what you don't want hard" still stands as the primary aspiration.

3) Participation Matters - Citizen involvement in planning is not a particularly dramatic notion these days. There are several reasons for seeking widespread participation in local planning efforts. Participation offers cover and legitimacy to decision makers and planners. However, it also contributes two essential elements that contribute to successful plans. First, widespread participation increases the range of ideas at the table. More eyes on the problem means more insight into its real nature and possible solutions (JPL article). Second, involving a wide range of folks makes ownership of the results, the plan, widespread, particularly during the years and decades that it takes to act on plans.

Planning is an important part of the kind of community building needed to sustain values and visions over the long haul. Participation made Metro's Region 2040 planning process possible. In the late 1980's Metro had no growth management program. It managed the region's urban growth boundary, but in a very limited way with one half-time planner. As the metropolitan area began to emerge from a prolonged recession during the 1980's, Metro was called on the review the urban growth boundary as part of the "periodic review" process in the Oregon statewide land use planning program. As part of that review, Metro proposed to develop a set of Regional Urban Growth Goals and Objectives, a long-deferred responsibility identified in its initial enabling legislation.

In the Portland way, a "Noah's Ark" of a committee was constituted to guide the process with membership consisting of elected officials from throughout the region, environmentalists, home builders, business interests, and citizens. After a year of developing goals for urban form in ten categories, the committee recognized that something needed to be done, and that Metro had the power to make it happen. The next year was spent developing the outlines of the regional planning partnership in place today, a component of the document that spelled out how Metro would do regional planning, and what the roles and responsibilities would be for all involved and effected.

At the end of this two-year process, it was suburban elected officials that observed that although they and Metro had developed a great description of the pieces of a well-functioning region, there was no vision, no overall description of where we were heading. It was the participants themselves that called for the creation of what is now known as the Region 2040 Growth Concept, not Metro. Metro's signature growth management planning effort would have never happened without the participation of a wide range of interests.

4) Leadership Matters - Unquestionably, Oregon and the Portland region have benefited from terrific leaders that have made critical contributions at different points in time. Former Governor Tom McCall and key legislative leaders from both parties made the Oregon statewide land use planning program a reality. Former Governor and Mayor of Portland Neil Goldschmidt has been widely credited with leading the revival of Portland's downtown, neighbourhoods, and civic culture. Today we continue to benefit from key leaders in critical positions, people able to accurately name the problem, see the connections between interests, and inspire us to be better than we are.

However, the role for leadership has changed substantially in the last 25 years. Leaders no longer control information like they used to. We have neighbourhood activists running ArcView on their home computers. Information is now everywhere. Furthermore, in this community interests are divided finely enough that we no longer find ourselves with one "lead" issue that pulls folks together. Consequently, the future for us will be in coalitions, and the leaders of the future will be those able to create the partnerships and collaborations needed to advance overall community values.

5) Good Things Take Time, or, Timing is Everything - Our region is a work in progress. It's not done and it never will be. Things weren't always like they are today. Lewis Mumford visited our region in the 1930's at the behest of a group of leaders investigating the ways that electrification could be integrated into the Pacific Northwest. Mumford had this to say about Portland and Seattle:

> ...neither Portland nor Seattle show, from the standpoint of planning, more than metropolitan ambitions that have overreached themselves. The melancholy plan to increase Portland's population from 300,000 to three million succeeded in disordering and unfocussing its growth: but it did little to give it the benefit of modern city planning practice; meanwhile, the apparent financial prospects of these port cities undermined the base of the sounder development that could well have been taking place in other parts of the region, on strictly modern lines.

His solution: build new towns in a landscape that literally took his breath away. Some 60 years later, downtown Portland is a national model, but downtown was being rebuilt in the depths of the recession of the 1980's because of initiatives put into motion 20 and 30 years before. The rebuilding of Downtown Portland, the removal of the Harbor Drive Expressway to create a public park, the revitalization of neighbourhoods and historic town centres, all of these things take time and can only occur when the timing is right.

Creating a great place, however you define it, and over whatever scale you are working at, is the legacy of a lifetime, not a matter of months, a single term of office, or the cutting of a single ribbon. Stewardship takes place over generations, not all at once. There are no single improvements, initiatives, plans, or programs that can, by themselves, change the course of urban development, redevelopment, and change.

6) You can't regulate Quality of Life into Existence - Planners have lots of tools for limiting damage. However, community quality of life is a collective achievement. The great parks of our region didn't result from exactions, but from bond measures and other collective measures. Modelling work accompanying the development of the Region 2040 Growth Concept found that even with the ideal arrangement of land uses and buildings with relation to the transportation system, transportation system goals would not be met without changes in trip making behaviour.

That is, absent a culture of inhabitation to go along with the physical development of the landscape, plan objectives would not be realized and community values wouldn't be served. Achieving quality of life goals is a partnership between what is required and the choices made by individuals. Regulation alone is not sufficient. Putting choice into context remains a central challenge for planners.

7) Things Change - There is a difference between planning when times are good and planning when times are bad. When times are bad communities seek change. When times are good, communities seek to stabilize the status quo, and planners speaking of change might as well be speaking in tongues. In the mid-1980's, our region sought any kind of change it could. In the early 1990's as the pace of growth began to pick up, the communities became more critical of the change that was occurring and the change that they sought. Today, there are a significant number of people calling for either no growth or slower rates of change, and today unemployment is at historic lows.

Recently the City of Portland engaged in a district planning process in the south western part of the city. The Southwest Community Plan was met with sustained and passionate opposition, especially when it became clear that the purpose of the plan was to change zoning to accommodate even more growth. However, the southwest district reported the highest level of satisfaction with neighbourhood conditions on a survey conducted at about the same time the planning effort encountered its peak opposition. In that case, neighbours weren't seeking change. When times are good, the challenge is to make the present work better, and planning strategies and objectives need to change. The only constant is the underlying values of the community.

8) Growth Management has Costs as well as Benefits - Planners have known for some time that sprawl is not free. Recent studies revisiting the "costs of sprawl" question in Oregon and elsewhere have reconfirmed that sprawl and its inefficiencies carry a cost. However, we've also learned that the alternatives to sprawl aren't free either. Every choice entails a cost, and growth management can make the nature of the challenge of creating equitable regions even more explicit.

Although numerous studies have been developed to determine the cost imposed by the urban growth boundary, none has been able to demonstrate that the urban growth boundary alone is the reason, even a significant reason, for rising housing costs in this region (Goodstein, OHCS, Florida). Nonetheless, choosing an urban growth boundary both incurs a cost of some magnitude and, perhaps most important, implies that traditional methods for supplying affordable housing--devaluing central cities and older suburbs while minimizing land costs elsewhere--won't be available here.

Our challenge, like that throughout the country, remains identifying ways to house the least affluent households in our region in locations close to services and jobs. Growth management doesn't relieve us of facing up to that challenge, just as sprawl doesn't relieve other regions of the same challenge.

9) An Urban Growth Boundary is not an Urban Growth Management Program - The motivation for the Region 2040 planning program grew out of the realization in 1989 that we were experiencing all of the same "sprawl effects" of other metropolitan areas within the urban growth boundary in this metropolitan area. On the other hand, it has been demonstrated that the urban growth boundary has protected farm land, helped to sustain farm production, and established a clear difference in urban and rural resource land values. Further, a parcel map of the region reveals that the urban growth boundary has clearly separated urban densities from rural densities.

Although the urban growth boundary has had the intended effect of protecting farm land and farming, it has not prevented sprawl patterns of development within the urban area. That is, it is a very powerful and useful tool, but no silver bullet. A comprehensive growth management program can benefit from an urban growth boundary, but it also requires action and attention at other scales.

Neighbourhoods, streets, town centres, main streets, regional centres, and green spaces all need specific attention. Design matters at every scale. High quality public spaces, creating locations of choice within the metropolitan area, don't happen by accident. Urban growth boundaries force conscious choices about urban structure and urban/rural relationships. What happens within the boundaries requires other tools and initiatives.

10) Community Building, both locally and regionally, is an Ongoing Responsibility - At the turn of the century, Portland's booster message was simple. Portland just wanted to be big, the "Queen City" of the Pacific coast. Today, on the cusp of a new century, a quest for bigness is not a sufficient booster message. Attention has turned to being better, and improving the quality of a place automatically brings into focus questions of cost, collective responsibility, and political will. Our experience is that regional planning is fundamentally a community building task. Furthermore, if people aren't empowered locally, if they don't feel effective in arenas close to home, they won't be able to relate to a regional plan.

Consequently, an ongoing effort needs to be made to build both strong local communities and collective recognition of a shared metropolitan future. One without the other won't ring true and won't go far. Ultimately competitiveness at a metropolitan scale is not just about doing things well that other places can do, but doing things well unique to your own metropolitan area. In Portland we live in a landscape with oceans, mountains, deserts, wilderness, and incredible fresh food all within the same day's drive. We can drink the water from the tap and still see the mountains on the horizon. If Portland can continue to make and remake itself as the best place it can be, then the strength of our collective ideas and action will continue to be of interest to the world. No one will visit us to see a better version of some other place.

Finally, this region has seen both growth and decline during the period that our notable regional and local planning efforts have unfolded. Significantly, two-thirds of the growth in population in this region has come from in-migration. Without a strong commitment to place, and efforts to continually build community, public support for initiatives that have taken decades to come to fruition would have shrivelled.

Conclusions

The fact that we can draw these lessons from our experience should not suggest that we've continued to live by and remember them in every situation. Portland, like every place, needs to continue to focus and refocus on our shared values, particularly as the community changes. Asking and re-asking questions about basic values is what keeps plans and communities alive. We are still struggling to understand both the impact of our choices and our responsibilities for addressing them. In some areas, like planning for the future accommodation of needs for air transportation services or linking land use plans across a state line, we are just now creating a common context for the hard work to come.

Growth management is primarily a game of rates: forecasting rates of growth, assigning growth to locations, attempting to match rates of growth with the provision of infrastructure and services. Growth management has been portrayed in the past as a means for accommodating new growth as if it never happened. That is, the impact of new growth would presumably be minimized as its requirements were recognized and addressed in advance through growth management efforts.

Though it is important to know how big things might get and by when, and to ensure that communities can grow into themselves gracefully when the time comes, we've learned that you can't manage away the effects of growth. More people means more activity, more competition for fishing holes, and more folks in the check-out line no matter how good the growth management effort has been. Growth, like decline, results in change.

There are several questions worth exploring through planning, but these are not typical growth management questions. First, how can new growth assist communities with maturing? Most often, discussions of growth are about adding something new. Rather than focussing on newness, how can growth assist places with becoming more mature, with better making the fit between local aspiration, environmental quality, and sustainability? Second, no matter how many people show up, and no matter when they get there, what ought to still be true about the community?

We've learned that planning cannot prevent change. Change is a constant. However, planning can advance values. What are the qualities that ought to still be evident, still be true about the place in the future? Planning can make great strides in that direction as evidenced in Portland's downtown, its neighbourhoods, and in the vitality of the working landscape surrounding the rapidly expanding "silicon forest".

Ultimately, as the "Eden at the end of the Oregon Trail", we are still trying to figure out just what a city in Eden looks like. Most of our planning to this point has been prompted by a desire to sustain Oregon's traditional economy, and its reliance on high quality farm and forest land. Many if not most of the innovations that we're known for are innovations of preservation. However, city building is a creative act, and an uncharted challenge in this country at a metropolitan scale. There is no blueprint for an urban vision in Eden. Our struggle in the years ahead will be creating for ourselves a vision of urbanity in this western place. Make no mistake, what we've done and what we're doing is all a work in progress. May it ever be so.

Chapter 15. Complex Urban Transport in Europe: Tackling Problems, Finding Solutions

Stephanos Anastasiadis

Introduction

European urban transport is an interesting case study. On the one hand Europe is frequently held up as an example to the rest of the world for its environmental thinking. On the other hand, urban transport developments are tending to move Europe towards less, not greater, sustainability. Much of this development has to do with the increasing dominance of the private car; in itself a marvellous tool, but overused in Europe's cities.

Transport is extremely complex. So many different factors govern its development that it is difficult to have and maintain an overview. Despite this – or perhaps because of it – responsibility for transport development appears to be fragmented, with clearly defined areas of expertise and not much communication between different disciplines. The whole complex picture is looked at very little and when it happens it is often treated with suspicion. For example, units dealing with road planning in city councils are not always known to deal with their colleagues in the public transport department as a matter of course, who in turn may not always communicate with the regional planning department. This is understandable, but does not help to achieve a higher quality of life.

This manuscript is based on the presentation given at the second Volvo Conference on Future Urban Transport, held in September 2003. However, it is being complemented several months after the event and time has not stood still. At the time of writing the EU had just grown from 15 to 25 member states, European Parliament elections were under way and the European Commissioners were in their final days in office. On the legal level, there was a chance that the new EU constitution would be agreed. In short, everything on the European level was once again open. A moment of opportunity was approaching.

A word of caution: European figures in this paper refer to the EU-15, unless stated otherwise. Although the EU has expanded to 25 countries, good aggregated figures are not yet available. In general, urban public transport has a greater market share in central and eastern European towns and cities, though this is rapidly changing.

Vision

A city that had sustainable transport would be one in which goods and services were easily accessible. The benefits of transport policies would be maximised and fairly distributed; the negative effects would be minimised and shared out fairly. It would be physically structured in such a way as to allow residents to move around freely, with easy access to green spaces. The borders of the city would be fixed so that horizontal expansion would be difficult. It would offer a clean local environment, with good air quality and a low noise level. It would be a place that people cared about and wanted to keep in good condition. It would be a cultural and economic magnet, a crucible of civilisation. Public transport coverage would be good across the whole city and the service would be affordable. Residents would be able to choose to use a car, but would not be obliged to do so in order to access such benefits as jobs and education. The city should be easy to enter and easy to leave. People using motorised transport would pay any costs this generated, which would otherwise have to be paid by the whole of society. The framework for the ideal city in Europe would be set partly at city level, partly at regional or national level and partly at European level.

Concepts

Effects of unsustainable urban transport

The environmental problems caused by unsustainable urban transport are well known and do not need repeating. However, there are two main social problems, which are less well known.

Unsustainable transport patterns increase social exclusion directly. The transport system over-privileges the private car, meaning that those who do not have access to a car are often at a disadvantage in accessing schools, jobs, the cheapest shops and so on. According to the UK government's social exclusion unit[26], poor transport can reinforce social exclusion.

The negative environmental effects of unsustainable transport disproportionately affect those already marginalised: the poor, the disabled, the elderly and children. This increases existing social exclusion. Such effects include (for example) the impact of pollution on human health and the mental and physical effects of noise pollution.

Barriers to urban transport sustainability

Of all the barriers to more sustainable urban transport, the lack of political commitment is particularly important. When the political will is lacking, the problems can be recognised, but are not deemed important enough – in practice – for there to be a real change in the systems in place. But political will is, of course, not the only obstacle.

The European Conference of Ministers of Transport (ECMT) has identified ten main barriers to creating sustainable urban transport systems[27].

---

[26] Making the connections: Transport and social exclusion. Final report published in February 2003. Available online at: http://www.socialexclusionunit.gov.uk/transport/transport.htm

[27] See "Implementing sustainable urban travel policies," ECMT, Paris, 2002; www.oecd.org/cem

They are:

1. Lack of a national policy framework for sustainable urban travel
2. Poor policy integration and co-ordination
3. Inefficient or counterproductive institutional roles and procedures
4. Public, lobby and press resistance to policies
5. An unsupportive legal or regulatory framework
6. Weaknesses in the pricing/fiscal framework
7. Misguided financing and investment flows
8. Analytical obstacles
9. Poor data quality and quantity
10. Wavering political commitment

Political commitment for change can arise only where there is clear perception of the need for change, and that it will happen soon. This is not presently the case. While politicians have clearly adopted the correct language, they – and large parts of their electorate – have not yet internalised the pressing need for change towards a transport system that is more sustainable. Solving environmental problems is still seen as a luxury and transport is not directly linked in the public mind with social exclusion.

It is possible to place too much emphasis on financial incentives and signals, but it is certainly clear that price signals that promote car use in cities, or which send confused signals, do not help matters.

The organisational structure is also a powerful factor: a simplified example will illustrate. The United Kingdom (outside London) has a completely liberalised public transport network. Bus companies are actively discouraged from integrated ticketing. This means that if I buy a ticket from one bus company I cannot use it to travel the same route on a bus operated by another company. I need a kind of database of companies and timetables in my head and to match my travel planning to the operator that most closely reflects my outward and return journey. This actively discourages people from using public transport and it is no surprise that public transport user numbers are falling. The public transport market has not been opened up to competition everywhere in Europe. Where it has, however, the structure has frequently been carefully controlled. It means that I can buy a ticket on a bus run by one company and use it anywhere in the city. I am an end-user, being able to board a bus wherever and whenever I like without having to worry what company logo it is displaying.

Bureaucratically, decisions about urban transport are not always easy to make. This is because there may be different levels of decision-making – from the municipality or borough to the regional level, and where a city is also a state, such as the German cities of Hamburg, Berlin and Bremen, the state level. All these different levels of responsibility can make it difficult to ascertain exactly who is responsible for what, and it is not uncommon for bureaucratic buck-passing to prevent decisions from being taken.

The European Commission is considering requiring an overall sustainable urban transport plan from cities. This would be excellent, as it would require co-operation between different levels and departments of government.

One of the key barriers to sustainable transport in cities lies in psychological structures.

The importance of individual behaviour and choice in both transport users and decision-makers cannot be stressed enough. According to the OECD[28], "travel behaviour is only marginally related to fundamental values and preferences. Rather, travel patterns and levels are more likely to result from a combination of *habits* and *circumstances*" (emphasis in original). A combination of poor public transport conditions and excellent conditions for car-use therefore pushes people who can afford it to use a car, creating a self-fulfilling prophecy and in the process further disempowering those without a car.

---

[28] (1999, page 4) "Social implications of EST". In, *OECD Working Party on Pollution Prevention and Control, Working Group on Transport, Project on Environmentally Sustainable Transport.* The Economic and Social implications of sustainable transport: Proceedings from the Ottawa workshop. Paris: OECD. ENV/EPOC/PPC/T(99)3/FINAL/REV1

The physical make-up of the local environment is a powerful factor informing people's first decision of how to get around: suburbanisation makes public transport less effective and more expensive and encourages car-use. Out-of town shopping centres frequently offer the lowest prices but are difficult to reach with public transport. In the city centre, the physical provision of a dedicated bus-lane will speed up bus-users' travel times and make timetable planning easier; while provision of free parking spaces will encourage people to drive. The provision of free parking space has been a major feature of shopping centres since the private car started becoming widely used.

Not everyone living in a city can drive a car: perhaps they are too young, too old or infirm, disabled, or don't have enough resources to finance a car. And not everyone who could have a car wants one. Drivers know they may get caught in congestion if they take the car, and they may well be aware of the environmental problems added to by their behaviour. So in a city with a decent public transport network, why do people choose to drive? Stradling, Meadows and Beatty[29] have identified two main reasons: identity/status and control. Public transport also frequently has an image problem, being seen as the 'mobility provider of last resort'.

Research into transport behaviour indicates that people's perceptions of different forms of transport are a large part of the problem. For example, research at Lisbon University on the choice between car and public transport (primarily buses) shows that people see different modes differently. These perceptions of different modes of transport are often not consistent with observable fact, yet they nevertheless influence people's actions vis-à-vis different transport modes. This reflects the situation in everyday life. Cigarettes, for example, are more threatening to individual human health than snakes or sharks, but they *appear* to be less dangerous as the effects are less immediate.

---

[29] *Identity and independence: Two dimensions of driver autonomy*. Paper presented at the "ICTTP 2000 International Conference on Traffic & Transport Psychology," Bern, September, 4-7 2000.

When evaluating public transport against the private car, an individual typically misjudges the experience of using a different mode of transport from what he/she uses. Car-drivers, for example, tend to believe that commuters taking a bus find the experience far more stressful than they actually do (though it does tend to be more stressful than using a private car); and those using the bus tend to think the car-driver is less stressed than he/she actually is. The perception of stress increases universally as congestion becomes worse, but private car-users consistently believe that public transport users are more stressed than they are, and public transport users believe that car-drivers are less stressed.

People use coping strategies to deal with stress. Unfortunately for those wishing to encourage more sustainable transport use, people find it easier to cope with stress when they are in a private car than in a bus. Public transport users therefore need to use comparatively more coping strategies to remain calm, though often they believe they do not need to cope with stress in congestion: coping strategies are often used at a pre-conscious level[30]. Coping takes up psychological energy, meaning that public transport users need to use more energy than private car users to maintain a similar level of calm.

Other factors in the choice of mode include:

- Comfort: car-users see comfort as extremely important, public transport users less so
- Price: public transport users tend to be more sensitive to price than private car users
- Perception of time: those using collective transport over-estimate the amount of time they take to travel, feeling that their travel time is much higher than it actually is: this gives them a higher level of felt stress
- Safety: most people think that public transport is less safe overall than private transport, though the reverse is actually true

[30] Meaning that people are not fully aware that they are employing coping strategies.

The most important factors in determining whether people will choose to use the bus are perceptions of stress and safety.

The quest for control, often equated with freedom, is one of the driving forces in contemporary society, and is a key component in people's attraction to the car. It is a pre-rational attraction, as public transport in fact often offers greater freedom in the city. This is exploited by groups such as advertisers by, for example, their placement of advertisements on bus-stop billboards, which specifically emphasise the sense of control that a private car gives.

Within the boundaries of their perceptions, people are typically acting entirely rationally in choosing to drive a private car instead of taking public transport. It is not surprising that people choose more polluting but apparently more hassle-free transport to get around the city. It is one of the factors which has undermined public transport specifically, and which prevents the build-up of enough political will to make real changes towards a system of sustainable transport. Politicians need to be brave to counteract received wisdom that a car is better.

Once someone has chosen to drive a car, the biggest factor maintaining this behaviour is force of habit and social patterns. Once people have chosen a way of meeting their mobility needs, they stick to it. Once established, patterns of individual and group behaviour change very slowly. This is unsurprising: there are so many different decisions to be made in a day that it is tiring to make them all consciously, so most decisions are made once and then become part of a routine, a stable pattern of behaviour. Changing this pattern takes energy and without strong impetus from outside it is unlikely that a car-driver will choose to take public transport (for example). This is a lesson for the EU's new member states, where public transport use is far higher than in the west: it will be far easier to keep the modal split than to attract customers back to public transport once they have made the switch to the car.

Just like people, social structures are also strongly resistant to change. This brings us back to the ECMT research that wavering political commitment is one of the hurdles to sustainable urban transport. Seen through the prism of transport psychological research, it is perhaps the most important barrier.

Access not mobility – small but significant difference

People make journeys to get to places: schools, shops, leisure facilities, work, holidays. In the vast majority of instances, particularly in cities, the motivation for travelling by car is to access the goods and services on offer at the destination. It seems that decision-makers conflate access and mobility, mistakenly believing that greater mobility will automatically provide greater access.

Planners often appear to understand planning for mobility to mean planning for motorised mobility. Yet this is only a partial solution. It provides incentives to use a private car, worsening pollution, increasing pressure for sprawl to develop and continuing to exclude people without the car. As long as this remains the case, transport cannot be sustainable. Instead, planning should aim to ensure access; 'accessibility-proofing' of policies being a useful concept. Such solutions may even prove to be less expensive than adhering to the myth that providing more mobility automatically improves quality of life.

Investment

Investing for sustainability clearly requires careful investment of money. Positively, this means spending on well-thought-out investments. Negatively, it means, doing no harm, not spending money on investments that are likely to harm the prospects for sustainability. In both cases, there needs to be clarity of goals – what the spending is actually for, what it is hoping to achieve – and then an analysis of how to achieve them.

However, investment means more than simply money. If money can be thought of as energy, as an enabling force, then it also makes sense to think of investment as expenditure of energy. In this case, investment also means effort and political will. The overall goal is to use resources responsibly.

Those making decisions on how to spend money therefore need to invest time, effort and research into examining the likely results of their expenditure. They also need to invest in communicating what they are thinking of doing and how the decisions are being taken. Without buy-in or at least acceptance by the most important affected parties, the effect of investing money will be significantly diminished – and it is possible that bad decisions will be made due to lack of information.

Most importantly of all, investment of political will is needed. A government – at whatever level – has a finite amount of political capital. How much will depend on the mandate it has, on how trusted it is and a range of other factors. A good example of this is the present UK Labour government: it had much political capital at the start of its mandate, but appears to have spent much of this on its actions on the international scene. While the government may have had the chance to address much-loved institutions in need of reform, such as the National Health Service, it is now almost inconceivable that the government will be able to make such a move. There is therefore a limit to the number of controversial issues a government can tackle – and because decisions on urban transport affect people's lives directly, they are almost always controversial. Controversial decisions can of course turn out to be successful, with the public then supporting them. In such a case, the political capital could be understood to have been repaid – one could argue that this is what happened in the case of London Mayor Ken Livingstone (see below). Nevertheless, a government must have the capital in the first place.

Vested interests

Those benefiting from a particular set of circumstances, no matter how beneficial or damaging for the greater good, will usually fight to keep their benefits. To use an example from outside the transport or environment field: the Canadian White Ribbon campaign is a campaign group of men working to combat violence against women, which for them means working to end the benefits that men receive over women simply for being men. They receive frequent media attention precisely because their call is so unusual: men voluntarily wanting to give up power. The power of vested interests should not be underestimated anywhere, from city level to the European scene.

European setting

Political context in Europe

Most Europeans live in towns and cities. In the 'old' Europe of 15 member states, the estimates ranged from 75% to 80% urbanised. The figure is lower in the new member states. One-fifth of the kilometres travelled in the EU are urban trips of less than 15 km and half of all car journeys are less than 5 km. The European Commission expects the total number of kilometres travelled in EU urban areas to be 40% higher in 2030 than in 1995 under a business-as-usual scenario.

And yet there has been little effective action on the European level over the past few years. This is because there has been a sort of paralysing conflict on the topic of transport and environment generally, and urban transport in particular, within the European Commission, which has prevented effective action. This is important for action on urban transport as the European Commission is the only EU institution with the right of initiative: only the Commission may propose European legislation. The paralysis has therefore blocked European action on urban transport.

A bit of background is important here for readers not immersed in the business of EU policymaking. The European Commission is effectively the EU's civil service. It is headed by Commissioners, who are effectively EU-level ministers. DG TREN is the directorate-general for transport and energy and is the European equivalent of a national transport ministry. DG Environment is responsible for environmental policy. A legislative proposal will typically be created as follows.

The directorate-general responsible for a particular field (TREN for transport) prepares legislation over a period of time, often in consultation with stakeholders. Once the proposal has been written, the DG will send it to other DGs for their opinions. This process, known as inter-service consultation, is not public and can result in major changes to a proposal. Once it has completed its process through inter-service consultation a proposal is ready to be made public. This is done when it is officially adopted as a Commission proposal by the College of Commissioners at their weekly meeting. The legislation is then ready to start its process through the European Parliament and the Council. The European Parliament is the EU's directly-elected body of representatives, while the Council represents the member states and changes according to the issue under discussion. Proposals on transport will be examined by the Transport Council, made up of EU transport ministers. The process then becomes quite complicated, though for our purposes it is not necessary to go into further detail.

The conflict within the Commission has been largely between directorates-general TREN and Environment. It has been partly an inter-institutional turf war, of the sort that is repeated across the world and which need not be revisited here, and partly genuine political disagreement.

This disagreement can be distilled into two opposing slogans: Treaty obligation vs Subsidiarity.

The Treaty establishing the European Community was agreed by all EU member states as the treaty under which all are bound. Articles 2 and 6 are the most relevant to the 'treaty obligation' camp. These state:

> The Community shall have as its task, by establishing a common market and an economic and monetary union and by implementing common policies or activities referred to in Articles 3 and 4, to promote throughout the Community a harmonious, balanced and sustainable development of economic activities, a high level of employment and of social protection, equality between men and women, sustainable and non-inflationary growth, a high degree of competitiveness and convergence of economic performance, a high level of protection and im-

> provement of the quality of the environment, the raising of the standard of living and quality of life, and economic and social cohesion and solidarity among Member States. (Art. 2)

and

> Environmental protection requirements must be integrated into the definition and implementation of the Community policies and activities referred to in Article 3 [one of which is transport], in particular with a view to promoting sustainable development. (Art. 6)

Given that the majority of Union citizens live in urban areas, that the Treaty requires action to ensure environmental protection and improve quality of life, and that environmental policies must be included in transport policies, the treaty obligation camp would say that the European Commission has the right and obligation to formulate policies that improve urban transport.

It therefore seems clear that the Commission does have the right to propose frameworks for urban transport which would improve quality of life in urban areas through (for example) improving safety, in much the same way as the Community has acted on compulsory seatbelts and motorcycle helmets across the Union. An example would be to determine maximum urban speed limits, which would have the added advantage of reducing air pollution and congestion.

On the other hand, the Treaty also contains within it the principle of subsidiarity.

> In areas which do not fall within its exclusive competence, the Community shall take action, in accordance with the principle of subsidiarity, only if and insofar as the objectives of the proposed action cannot be sufficiently achieved by the Member States and can therefore, by reason of the scale or effects of the proposed action, be better achieved by the Community. Any action by the Community shall not go beyond what is necessary to achieve the objectives of this Treaty. (Art. 5)

The Commission defines the principle of subsidiarity as follows:

> The subsidiarity principle is intended to ensure that decisions are taken as closely as possible to the citizen and that constant checks are made as to whether action at Community level is justified in the light of the possibilities available at national, regional or local level. Specifically, it is the principle whereby the Union does not take action (except in the areas which fall within its exclusive competence) unless it is more effective than action taken at national, regional or local level. It is closely bound up with the principles of proportionality and necessity, which require that any action by the Union should not go beyond what is necessary to achieve the objectives of the Treaty[31]

Generally speaking, the Community has clear competence to propose EU legislation on market-driven concerns and cross-border issues, such as rail liberalisation, air quality standards, minimum standards for public service requirements and award of public service contracts in public transport. The Community also has competence on technical standards.

For obvious reasons, the Community has least scope to act in issues affecting only the local level. These include, for example, decisions on specific urban spatial development, such as where to put bus lanes.

Within these two extremes, the limits on Community competence under subsidiarity is up for debate and subject to political discussion.

In practice, the decision of whether or not to propose EU legislation on urban transport or whether the principle of subsidiarity trumps it is a strongly political question.
At present, the Community does not take a leading role in urban transport and there is a case for the argument that the Commission can and should do nothing other than fund good projects and support

[31] Glossary of EU terms on the Commission website:
http://europa.eu.int/scadplus/leg/en/cig/g4000s.htm#s10

best practice. However, the Common Transport Policy White Paper[32] does have a short section on urban transport, which Commission officials say means that the EU now considers urban transport as part of its transport policy.

The Community also has the competence to provide legislative frameworks for urban transport when competition is at stake – particularly in border regions. The provision of parking spaces, for example, could easily be a competition issue: if one city does not provide free car parking for a business wishing to move there, the business can go to another city which does provide the amenities[33]. There is thus an argument to justify the EU setting the framework: the details would then be up to the regions themselves.

On the other hand, it could be argued that cities should be free to decide the conditions under which they operate and that in a democratic process the will of the people affected by a decision is the highest arbiter and should be respected. This argument certainly has merit, and in a perfect world would be irrefutable. However, it ignores the very real pressures on decision-makers to provide short-term gains, particularly employment, possibly resulting in decisions that are not taken on the basis of full or even good knowledge. City representatives have often pointed out that the desire to provide jobs will frequently override all other concerns.

[32] The Commission White Paper on the Future of the Common Transport Policy (CTP) was adopted in September 2001.

[33] The point can be illustrated by referring to airport noise. The freight delivery company TNT used to be based at Cologne Airport. When the authorities refused permission to TNT to expand its activities, the firm looked for another European base. Many airports competed to offer TNT the most favourable possible conditions. Monetary and legislative incentives were offered, the decisions being taken as quickly as possible to ensure attracting the putative benefits of TNT's business. Decisions were taken without a cost/benefit analysis. TNT moved to Liege Airport in 1998. It was immediately evident that the noise costs were very high and it has since emerged that the health costs from the extra noise pollution are heavier for Liege than the economic gains. Regional authorities in Strasbourg resisted the temptation to jump at the short-term gains of allowing the freight company DHL to operate at Strasbourg Airport. Instead, it commissioned a cost-benefit study, on the basis of which it refused access to DHL. (from: Noise in Europe (2000), an NGO community briefing, available on the T&E website, www.t-e.nu, under Activities\Health\Position papers)

Member states are very nervous about allowing the Commission to set frameworks for urban transport; the reason they give is subsidiarity. It is already clear that general frameworks, such as the air quality daughter directives[34], will force cities and regions to take action and that goes quite far enough for many member states.

Subsidiarity is a complex issue, and it is no surprise that the Commission is hesitant to act, despite its obligations under the Treaty. Ultimately, the question of whether to take action on the European level, or whether subsidiarity is a reason for inaction, is a political one. As one Commission official put it in 2002: whether or not the Commission will take a particular action rather depends on the 'way the political wind is blowing. As another pointed out, subsidiarity is 'often used as a smokescreen,' a convenient way of avoiding politically difficult decisions.

This political reality makes the arguments about subsidiarity fade into the background a little. The Commission will not propose something for which political will is lacking, meaning that it will not use its right of initiative to propose legislation which has a direct bearing on urban transport[35].

As of 2004 the discussion is increasingly of the 'need for competitiveness', which also has a mention in the Treaty.

Ultimately, the conflict has been based on the myth that environmental concerns are in conflict with economic development, a powerful argument in times of economic hardship. It strengthens the hand of those wanting to leave responsibility for urban transport entirely with cities. Yet a wide range of creative solutions is needed,

---

[34] These come from the Framework directive on ambient air quality management and assessment (96/62/EC).

[35] For example, the Commission tabled environmental legislation in the early 1990s on the back of a wave of political will which culminated in the Rio summit – before there was an explicit legal basis in the Treaty for environmental legislation. There has in fact only been an explicit basis for environmental legislation since the Treaty of Maastricht came into force in 1992. Later attempts to generate environmental legislation under that treaty have met with much more resistance, despite the legal base, because the political will to take action had lessened.

by all actors if the problems caused by present urban transport patterns are to be solved.

DG TREN has principal responsibility for the Commission's transport policies. Its mantra has effectively been, 'Demonstrate, not legislate' in the field of urban transport; for the reasons mentioned above. One example of the demonstration approach is the Civitas project, a €50 million demonstration project to support "ambitious cities in introducing and testing bold and innovative measures to radically improve urban transport". The CUTE project is another example, demonstrating fuel cell buses in a handful of European cities.

However, there has been little environmental progress in the transport field. The European Environment Agency, an official EU agency, issues an annual set of indicators called TERM, the transport and environment reporting mechanism. These indicate that the environmental performance of transport is hardly improving at all. Eurostat figures show that transport's CO_2 emissions are constantly increasing and are the fastest-growing sector in Europe. This growth in emissions more than cancels out energy-efficiency gains made by industry. Road transport emissions are growing at 2% a year. The environmental provisions in the 2001 Common Transport Policy are not being respected[36]. And the 1999 transport integration strategy – designed by EU transport ministers to meet the integration requirement of Treaty Article 6 – has yet to be put into effect. In short, not much is happening and transport continues to weigh heavily on Europe.

One explanation for this lack of progress is that it is not possible to solve a social problem with technology. Demonstration projects are basically about using technology as fully as possible. While technology is important, it is not the way to solve the problems of urban transport.

DG Environment has taken a different approach in an attempt to create the essential political will.

[36] Although this is not directly relevant only to urban transport, it is worth mentioning as the CTP is supposed to be the blueprint for transport policy over a decade.

The EU has developed a Sustainable Development Strategy, agreed by EU heads of state and government. The environmental component of the SDS is the 6th Environment Action Programme, run by DG Environment. The 6EAP is split into seven thematic strategies, one of which is a thematic strategy on the urban environment. The urban environment thematic strategy has four priority themes, one of which is sustainable urban transport. The work on urban transport under the 6EAP is an attempt to look at urban transport in a coherent and inclusive way, taking complexity into account. Although DG Environment is not officially responsible for transport policy, no strategy on the urban environment could reasonably ignore transport, so it has been allowed to include it, with input from DG TREN.

The urban environment thematic strategy (reminder: one of seven thematic strategies under the 6EAP) is to take account of, among other things, the following: consider urban environment indicators; the need to tackle rising volumes of traffic and bring about a significant decoupling of transport growth and GDP growth; a reduction in the link between economic growth and passenger transport demand; the need for an increased share in public transport, rail, inland waterways, walking and cycling modes.

In other words, the work on urban transport is to go beyond technological fixes and look at demand management.

DG Environment is acting cautiously, aiming to slowly develop political will for action on urban transport on the European level on the basis of a 'knowledge-based approach.' Step 1 was to understand the problems. This was designed to develop an *état de lieu*, an overview of the problems of urban transport that involved all stakeholders. Thanks to the work done by bodies such as the EEA, we already had a fairly good idea of the problems, but DG Environment, not unreasonably, wanted to ensure buy-in from all stakeholders and this necessitated a longer process.

Step 2 is now to develop solutions, timetables and precise measures. The latest step in this process was taken in February 2004, with the publication of the Communication (discussion paper), "Towards a thematic strategy on the urban environment". A broad consultation exercise is now ongoing, with meetings until the end of 2004, and the final version of the thematic strategy is expected to be published in early 2005.

This is an excellent approach, because it takes complexity into account and looks beyond technology. It could lead to European level legislation on urban transport, which is what the thematic strategies are meant to do. However, the debate between the 'treaty obligation' and 'subsidiarity' approaches has not been resolved. The process followed in the DG Environment approach may increase the pressure for European legislation, but in the end whether it works or not will strongly depend on which way the political wind is blowing when it emerges.

'Back door' legislation

Some European laws will have a large and positive impact on urban transport development, although they were not designed for this purpose. There are numerous examples, but two will suffice here.

The air quality framework directive, adopted in 1996, has spawned so-called daughter directives governing emissions of certain pollutants. The first daughter directive is the most relevant for our purposes and is already in force. It sets standards for emissions of SO_2, NOx, particulates (PM_{10}) and lead, the aim being to improve air quality and thereby protect human health. The limits must be attained within a specific deadline. In cases where the concentrations of pollutants are very high now, member states must prepare action plans showing how they will go about achieving the limit values. Although the daughter directive is already in force, there is a so-called margin of tolerance, which allows limits to be exceeded. It is a percentage of the limit value and differs for each pollutant. Over time this margin of tolerance decreases to zero, which will be the absolute limit with no exceptions. Depending on the pollutant this is 2005 or 2010. The first daughter directive establishes the exact principles for monitoring each pollutant and requires clear reporting to the public of limit values and air pollution levels. Although monitoring is foreseen as a national responsibility, planning and informing the public are principally thought of as local and regional tasks.

It seems very likely that no city in Europe will meet its air quality targets under business-as-usual scenarios. Cities and national governments are slowly starting to realise that meeting the air quality standards will be a challenge, and – importantly for urban transport – will require changes to the way transport in cities is structured. Technology cannot work on its own: demand management will be needed. This has led to discussions in Brussels about weakening these air quality standards. That would be unacceptable, and it would in any case take several years to change the legislation. In the meantime, the European Commission can be expected to start calling member states to book for failing to meet their commitments. Although it is rare for a member state to continue to fail to comply with legislation once the Commission has opened infringement proceedings against it, in the end this process can lead to substantial daily fines. We can therefore expect national governments to put pressure on city authorities to meet the air quality targets. This is in addition to any local political pressure that may come from residents who have access to information on their local air quality.

For this reason, the air quality framework directive could have a strong and positive impact on urban transport development.

The second piece of backdoor legislation has not yet come into force and is indeed still under discussion, because of it being extremely controversial. There is no need for a lengthy discussion of the Commission's proposed Regulation on public service requirements in opening up public transport markets to controlled competition. The Commission is convinced on the basis of research that opening up public transport markets to controlled competition can lead to more efficient and more popular public transport services. The total transport market in the EU is worth €95 billion, of which €24 billion is already open to competition in some form or another. The Commission proposed legislation in July 2000 to provide legal certainty in public transport markets. After much heated discussion, the Commission took the unusual step of withdrawing its proposal, and put forward a new one in February 2002. There has still not been much progress.

The point of this example is three-fold: firstly, if actually adopted, the legislation would set minimum environmental and social standards which must to be met in public transport – which could go some way to improving local urban transport and improving quality of life. While authorities putting out public transport to tender can already demand strong social and environmental standards, it is not an automatic requirement. Secondly, the delay in the legislation's progress is indicative: the lack of agreement comes less from the fact that the legislation is complex (although it is), than from the fact that the discussion is taking place on both the practical and the ideological level. Although the legislation has, on closer inspection, little to do with traditional left-right politics, it has been reduced to this in political discussions. Thirdly, it shows the limits of relying exclusively on legislation from Brussels to solve problems.

Local and regional setting

This is principally a paper about the European level, so it will not go into any detail about actions on other levels of responsibility. Having said that, one good example stands out: demand management in London through the now-famous congestion charge.

Briefly, the charge started in February 2003. Cars entering the charging area (which roughly corresponds to the city centre) must pay a daily charge of £5 (about €7.50). Payment is easy and can be done in many ways. Failure to pay leads to heavy fines and ultimately prosecution. Not all road-users are required to pay – cyclists, motorcyclists and certain categories of car-users are exempt from the charge – and residents within the zone pay a heavily discounted fee. There are many criticisms that could be levelled against the charging scheme. However, there is no doubt that the charge has been effective: fewer vehicles enter the zone; traffic within the charging zone is down significantly, leading to smoother and faster travel times; and public transport use is up. A MORI poll four months after the scheme started indicated that 50% of all Londoners were already in favour of the charge, while only 34% were against, with nearly three-quarters of respondents saying the charge was working. Most importantly, London's mayor, Ken Livingstone, who introduced the charging scheme, was re-elected as this paper was being written, well over a year after the scheme was introduced, providing a definitive indication of public support.

Why has the congestion charge been a success? Put in simplistic terms, it is because it has been accepted by the public. In addition to the fact that the traffic situation in London had been considered dire, and there had been general agreement that 'something had to be done', two key factors played a role, both related to how the charge was created. Firstly, the charging scheme has been communicated excellently: there has been good public information and much public debate on the matter. Secondly, the revenues have been earmarked for public transport improvement: this is important because clarity on the use of the revenues is essential for engendering trust in the public and public transport spending is seen as socially progressive.

Pricing alone will not unglue drivers from their private cars. It will take a wide-scale change in perceptions, encouraged by a real improvement in available options. Other solutions include good land-use planning, investment in public transport and the infrastructure for cyclists and pedestrians, and a focus on providing access rather than mobility. These elements are all present in the London congestion charge. It remains to be seen whether the change in perceptions is permanent.

It seems clear that Livingstone's willingness to take a political risk makes the London scheme stand out from approaches aimed at tackling urban transport problems on the European level. It is far more appropriate that such initiatives are taken on the city level.

What Europe needs to do, however, is create the framework within which it is possible to take necessary steps and exert pressure to do so through requiring well-communicated sustainability planning.

Europe needs Livingstone's clarity of goals, even if it does not and should not require congestion pricing in every city.

In Europe, the approach is significantly more risk-averse. This is partly because there are so many more stakeholders involved and partly because there has been a culture in which it is acceptable for the Commission to avoid responsibility for its areas of competence. Under the circumstances, the present work under the 6EAP is probably about as far as DG Environment can push the Commission.

Private practice

Government is certainly not the only actor in creating sustainable cities. There is also scope for private companies to make a big difference – either by investing in new business opportunities or investing in an effort to avoid poor investment in the first place.

An example of the former is car-sharing provision. This approach has been pioneered in Switzerland, where Mobility Switzerland provides what is effectively short-term car hire for people who wish to have use of a car without owning one. It makes economic sense for users driving less than about 10,000 km per year. The company has formed alliances with public transport operators and the national railways so that Mobility members get their annual transport passes at a reduced rate and the car-shared cars are placed at railway stations. Mobility has worked with large supermarkets to ensure that car sharing is well publicised. The keywords are good co-operation and communication, helping to ensure that users of the scheme have access to a vehicle when needed and access to public transport otherwise, all as a means of accessing more efficiently the goods and services that are the point of travelling in a city. Importantly, car-sharing schemes help to break the link between car ownership and identity, one of the most powerful pull factors for car ownership.

The car-sharing concept has spread somewhat. In Belgium the scheme is called Cambio and is being partly run by an NGO[37], in a direct effort to make concrete changes on the ground.

Another example is from a large corporation, Pfizer. The pharmaceuticals company said in early 2004 that it was paying employees to leave their cars at home. Staff parking at the firm's UK centre in Sandwich cost Pfizer €9 per day, so by offering up to €7 per day to walk, cycle or take public transport, the company has been able to cut car use by 15% as well as save around €145 000 so far. This is a clear example of a company investing time and effort in order to save money.

Conclusions

We understand well enough the problems. We understand well enough a series of individual tools. The challenge lies in putting the tools together to develop a way forward. The biggest challenge of all is in breaking patterns – patterns of belief (myths) and patterns of behaviour.

There is no single 'magic pill' that cities could take to somehow make urban transport sustainable. Solutions are multi-layered and complex. They require coherence between different levels of government and different departments within administrations. Political will is essential. The need for debate is equally important: those who are most affected, whose daily lives are presently affected and whose lives will be affected in the future, must have a chance to have their say.

This paper has indicated that there are potential, partial solutions at all levels, from the European level to the individual company. All need political will to implement and in some cases, political will is needed even to contemplate ideas (controlled competition in public transport, for example).

[37] The NGO is Taxistop, www.taxistop.be

On the European level, the EU has a responsibility to ensure equal health and social protection for citizens – through air quality legislation for example – but it is also responsible for ensuring equal conditions between cities and regions to avoid a 'race to the bottom' – particularly in economically difficult times. The exact degree to which 'Europe' can determine city policy should rightly be left to the political process to define: no matter how frustrating democratic processes often are, the alternatives are far worse. Personally, I think it is perfectly acceptable for the EU to require cities to develop action plans for sustainable transport

More importantly, Europe has huge coffers and gives an enormous amount of money to a range of projects. Every cent it gives or lends that leads away from sustainability is a wasted cent. Europe has a clear responsibility to examine its giving and lending criteria and avoid situations where it is simultaneously providing money for good public transport technology (good sustainability balance) and helping to finance a motorway that will increase the level of traffic in and around cities while speeding up the process of sub-urbanisation. Even if this means the decision-making process is longer, in the mid-to-long-term this can only be a good thing.

Whatever Europe does has an immense impact on cities, where most Europeans live. It therefore needs to ensure that at the very least it 'does no harm'. This means it should stop sending mixed messages, supporting pollution-increasing activities on the one hand and passing pollution-restriction rules on the other. Coherence is absolutely essential: you can improve public transport as much as you like, but if there is major motorway and road building going on in the same area, urban transport will remain problematic.

A wide range of creative solutions is needed by all actors if the problems caused by present urban transport patterns are to be solved.

This mix of measures must include:

- Transport pricing, with socially just use of revenues

- Investment in public transport
- Good land-use planning
- Shifting the focus from mobility provision to guaranteeing access
- Most importantly, a legislative framework is needed within which these changes can happen

The basic formula for sustainable urban transport in Europe is as follows: provide citizens with access to goods and services, while encouraging behaviour change by rewarding or punishing transport behaviour on the basis of its environmental, economic and social consequences. To implement this formula will take political will and good planning. On the European level, the Commission cannot legislate on all of the above, but it also cannot deny its responsibility to do more than simply promote good practice. The 6th environment action programme is promising, but we shall have to wait and see what comes out in 2005. NGOs have a responsibility to maintain pressure on their authorities to create good frameworks.

Chapter 16. Economic Impact of Light Rail Investments: Summary of the Results for 15 Urban Areas in France, Germany, UK and North America

Carmen Hass-Klau and Graham Crampton

Introduction

This present study analyses the effects of light rail investment in 15 urban transport areas. Five were located in France, four in Germany, three in the UK, two in the US and one in Canada. Economic effects are very difficult to attribute solely to one cause, such as the opening of a new light rail line. In virtually all cases other events occur simultaneously which also affect the outcome. Methodologies to isolate the effects will normally rely on data that show sufficient independent variation of the various influences, such as combined time-series/cross-section data in many areas over a long period of time, or the use of control cities and areas which are alike in every respect except the investment considered. All of these methodologies have their weaknesses and some of them are simply not practical in an empirical context. We have not developed a new methodology; we conducted discussions and interviews in each transport area with selected stakeholders, representatives of interest groups and some occupiers. These included public transport operators, the local government organisations in the cities and districts, business representatives and developers. In total, we interviewed over 70 experts. In addition, we visited all the new light rail lines and made notes about any new housing, office or retailing development close to the routes and stops.

The population size of the cities and transport areas selected varied from about 200,000 to 2.6 million; the light rail track networks we researched were between 14 km and 117 km in length. The number of annual light rail passengers per track-km varied from 240,000 to 1.7 million. The cities include examples of city centre pedestrian areas from 0 to 9 kilometres of streets; from 4,000 to 24,000 parking spaces in the centre; and from 400 to 9,000 park-and-ride spaces. These large variations show the complexity of the urban areas but also the different policy approach.

We divided the economic effects into:

- direct indicators (values of properties and rents close to light rail stations)
- indirect indicators (pedestrian flows in the city centre, reduction in car ownership and car parking spaces, economic benefits for employers and the workforce, overall economic gains in city centres and urban sub-centres)
- land use indicators which are not directly economic (change of retail character in the city centre and in older industrial areas)

Direct indicators

Only in some cities could we collect facts and figures. The easiest values to obtain estimates of seemed to be the increase or decline in residential property values.

Residential property prices

In the British city of Newcastle on Tyne the property prices close to light rail stations were estimated by estate agents to be 20% higher than similar property without light rail access. In Portland house prices close to light rail stations were 10% higher than further away. In Strasbourg a 7% and in Hannover a 5% rent increase is mentioned for the light rail lines. It was pointed out that house prices were higher around light rail stops in Nantes but not by very much. In Freiburg the rent was slightly higher in houses with good access to tram stops. House prices rose in Rouen by about 10% in close proximity to light rail stations but they also increased along the new TEOR busways. There were also higher house prices along the Altrincham line in Manchester and estate agents gave estimates in the order of 10% but emphasised that good access often had its main effect in quick sales. In Montpellier very expensive housing has been and is being built next to the tramline. No price increase was mentioned in five of the urban areas (Saarbrücken, Orléans, Birmingham-Wolverhampton, Calgary and Pittsburgh). In contrast to the other French cities, in Orléans there was a considerable decline in house prices during construction. This had been reported in a previous study about Sheffield (Chapter 2). Table 16.1 summarises these assessments, though it should be remembered that in some cases they are based on local judgements and estimates.

City	Residential property price differential in the neighbourhood of public transport improvement
Newcastle upon Tyne (house	+20%
prices)	+10%
Greater Manchester	+10%
Portland (house prices)	> 5%
Portland Gresham (rent)	+7%
Strasbourg (rent)	+10%
Rouen (rent and houses)	+5%
Hannover (rent)	+3%
Freiburg LRT stops (rent)	Expensive housing
Montpellier (property)	None – at the beginning nega-
Orleans (apartment)	tive because of noise
	Small increase
Nantes	No figure known
Bremen	None - at the beginning nega-
Saarbrücken	tive because of noise
	No figure known
Birmingham –Wolverhampton	No figure known
Calgary	No figure known
Pittsburgh	

Table 16.1. Change in property prices or rent at light rail stations

Office rent

Data on office prices were far more difficult to quantify. Most of the cities studied did not provide any differentiation in rents. Office rents in Strasbourg were 10-15% higher compared with other French cities of a similar size – Strasbourg's prices were in fact close to those in Paris.

In Freiburg, office rent on the periphery with very good road access was nearly 37% per sqm lower than offices at similar locations with direct tram access. A comparison of two office blocks, which were built at the same time and with the same quality, shows that the offices with tram access have 15-20% higher rents than offices without, even though these offices are closer to the city centre.

In Bremen we studied seven different areas zoned for commercial and office use. Although there are always other factors that may be even more powerful than public transport accessibility, the fact remains that all those areas which have good light rail access, except one, had 50% higher land prices compared to those sites which have no public transport or only bus access.

In Nantes we were told that commercial property prices are more expensive along tram corridors. About 25% of all new offices built in Nantes were located along a light rail line. In all European cities office location next to light rail was important. In Hannover it was pointed out that offices not on a light rail line could hardly be rented out at all.

Indirect indicators

Pedestrian flows in the city centre

With the opening of new tram stations in a city centre the number of *new* shoppers attracted to a town centre can be increased substantially. We obtained very good data on pedestrian counts in the city centre of Strasbourg.

On a Saturday in February 1992 (before tram Line A had opened) 88,000 people were counted at 11 locations in the city centre. The number increased to 146,000 in October 1995, about one year after Line A had started running through the city centre (an increase of about 66%). Certainly not all of these pedestrians bought something, but the size of the change is so substantial that it must have had an effect on retailing turnover in the city centre. The number of pedestrians increased steadily up to 1997 but then declined.

When Line B in Strasbourg opened in 2000 one could see a further increase in the number of pedestrians from around 121,000 in October 1999 to 160,000 in November 2000 (an increase of 32%). In the following years, the number of pedestrians fell slightly, though still remaining higher than before the two tramlines had opened. In October 2002 there were still 47% (41,000) more pedestrians in the city centre on a Saturday than in 1992. It is not clear why the decline occurred – one suggestion is that the increase was caused by a temporary 'novelty' effect of the new trams. A more likely reason is that during the last couple of years Strasbourg has been suffering from the German recession. Many Germans liked to come to Strasbourg to do their shopping and breathe a bit of French atmosphere: these visits have fallen off significantly. Unfortunately, there was no other city or transport area that could provide similar data.

A different economic effect of the new light rail line is seen in a medium-sized town in Germany (Saarbrücken). The line runs in the city centre, but parallel to the main pedestrianised shopping streets; it is not viewed positively by the representatives of the retailing organisation. However, the fact is that the number of public transport passengers did increase with the new light rail line from 25,000 passengers each day in 1997 to 41,000 in 2002. The overall *impression* from nearly all of the interviewees was that retailing, even in the town centre, has improved with this new mode of public transport mode but we have no proof of this.

In Germany and France the retailing sector wants direct access to trams and this is most important in the city centre. Of interest is the example in Bremen, where there were plans to move the trams out of the shopping street 'Obernstrasse' into the parallel street 'Martini

Strasse'. This would have moved the trams further away from the shops and the plan was not put into practice because of the protests from city centre retailers. Even for edge of town shopping centres, trams are important but this could have a negative effect on the city centre if shopping there is not attractive enough.

Car ownership

An interesting question is whether access to a tram reduces car ownership or the growth of car ownership. In order to find out whether this is true or not we looked at French and British census data for our eight case study areas and we also included Freiburg and Bremen in our statistical analysis.

One of the aspirations of the modern tram as a high-quality congestion-beating mode of public transport is to offer middle-class professional households the option of 'keeping their car ownership down' to one car, when they could well afford two cars. If this can be seen to occur, then it constitutes another important economic impact of light rail, which has hitherto been discussed very little.

In all our case study cities car ownership is lower in the tram corridors (300m and 600m) than outside. When studying changes in car ownership over time, the French and German cities showed some striking evidence of the ability of good light rail access to restrain car ownership growth, especially for multi-car households. The British cities are rather weaker in this respect.

Freiburg (Germany) is the most remarkable case, where car ownership has fallen less inside the 300m light rail corridor (5.8%) than outside (6.4%) whereas the number of households increased by 5% between 1995-2002. There was no significant difference in the fall in the number of cars between the 300m and the 600m corridors. The cars/household ratio in 2002 inside the 300m (600m) corridors was 13% (14%) lower compared with outside.

Economic gain in the city centre and economic sub-centres

In nearly all transport regions where a new tram or other rail infrastructure is being built there is the issue of who wins and who loses from this new accessibility. It is mostly retailers in the large city centers who seem to gain but small towns along the tramline are worried that they will lose trade. The fact that large city centers gain is already known from other studies e.g. San Francisco, Vancouver, San Diego, Düsseldorf . The effect on smaller towns and sub-centers is hardly known in any detail. Our interviews suggest strongly that both gain. Sometimes, as in the case of Portland, it could even be said that small towns gain proportionally more, particularly if there is sufficient land available for new development, reinforced by initiatives and political support from the city itself. Other cases where this fear was unfounded include the small French town of Saargemünd and the sub-centers along the new light rail line in the Saarbrücken region.

This is also true in Britain. Not only have the city centers in the large agglomerations gained economically, but the sub-centers and small towns have also benefited. Because of the fears that smaller towns would lose out when a light rail line is built, a number of these sub-centers have been pedestrianised and received money from Government bodies to improve their town centers and have done well.

No negative response to light rail was evident in any of the urban areas. The example of a small town near Strasbourg is interesting. Because of protests from the traders the alignment had to be changed, bypassing the town centre. However, it may well be that the traders will regret this decision, as was the case in another small town near Strasbourg, which protested successfully years ago against a tram stop. Today they are annoyed that they have missed the chance of direct tram access.

Reduction in the number of parking spaces

Offices and residential properties in close proximity to tram and light rail stops do not need to provide as many car parking spaces as the same building in a suburban area without good public transport access. Large car parking requirements will increase the construction costs for offices and housing and could reduce the total rent per square metre of land (if the surface space is used for parking). This is to some extent also true for underground parking.

Developers benefit if a large proportion of the workforce can travel by public transport as the full amount of car parking space is not required. We found in nearly all the study cities some reduction in car parking spaces along light rail corridors. According to the city car parking regulations in Bremen no car parking facilities at all need to be provided at tram stops. Thus a reduction in car parking is a substantial saving for the developer. In Germany it brings revenue to the city if a developer wishes to reduce the number of car parking spaces in excess of the regulations (which allow for generous car parking). For each car parking space omitted, the construction cost of the space must be paid to the city.

Employers

Good public transport access for the workforce is crucial to most employers. In many cases this may be the deciding factor for where a business will be located, especially if the workforce is low wage. Even in countries like the US, where the car is the dominant mode of transport, business decisions are sometimes based on good public transport access. We were told that the location decision in the city centre of Pittsburgh by PNC and Mellon, two large financial headquarters, was made because of the proximity of the light rail stations. In Nantes several large employers settled in the city centre after light rail line 2 opened, providing good access to the centre. The location concentration of offices in the city centre of Calgary can be largely explained by its excellent accessibility by light rail. About 80% of the workforce of the large Sears department store, for example, commute by light rail.

Another argument which was often heard during our interviews is the improved reliability and productivity of the workforce when arriving by public transport, as the stress level is much lower than when travelling by car. This has in recent years become more important with increasing urban congestion.

Workforce

Where car-parking fees are charged during office hours, they are in the majority of cases more expensive than public transport fares. Even in the US cities a large proportion of the labour force working in city centres cannot afford car-parking charges. This was certainly true in most cities, even in North America. Thus using public transport instead of the car benefits household budgets. The lower level of car ownership found in the tram corridors reinforces this argument.

Land use indicators

Change in retailing structure

Large rent increases for offices and retailing in the city centre as a result of a new tramline can change the retailing character in some streets from an average type of shop to high-class fashion and shoe shops. This change occurred in the city centre of Strasbourg when Line B opened. In Strasbourg retailing turnover increased in the city centre so much that the resulting increases in property prices and rents had a negative effect on the pattern of shops, driving the old-established shops out of the more expensive city centre streets.

Change in character of older industrial areas

Access to a tram station (or tram line) may also change the character of an industrial area. Such an area becomes more attractive for the tertiary sector, leisure and cultural activities. When one activity starts to locate there, others follow, although after a short while it is not clear whether new activities locate there because of existing activities or because of the tram access (it is possible that both factors interact). We found several examples of such changes in Bremen, Freiburg, Rouen and Strasbourg. However, when studying the German town of Saarbrücken, which has plenty of old industrial areas and is suffering from the current German recession, the new light rail access does nothing to attract offices there.

Lessons to be learned

In all the countries we studied one important common problem is that transport agencies and operators have little knowledge of the best alignments for development. This is one of the reasons why economic development along light rail lines is not as successful as it could be. In some cities, in particular in Germany, there was no policy to promote transit-orientated development simply because the current economic climate was seen to be too weak.

The key lessons to be learned are the following:

a) When land use and transport planning are combined – high-density offices and/or housing – often upmarket housing - will be constructed. We saw this in all countries we researched, but without active initiatives very little is happening. An active public sector is necessary.

b) The proposed alignment must go where development is taking place or will take place – and not where it is cheap to build a light rail line. That can occur as an exception, but such construction could create negative attitudes towards light rail within the community and may make further extension or new lines more difficult. Co-operation with private developers as well as with city and regional planning agencies is crucial when choosing a new alignment.

c) Any funding of public transport must take the potential for development into account. A line must be successful in two ways: in terms of passenger gains and in terms of attracting new property development. If such new developments can be assessed positively, funding from private sources will be easier. There must be flexible financial measures, such as tax increment financing, which make it easier to support light rail developments.

d) A good marketing campaign by politicians and the public transport operator, before, during and after the construction of a new light rail line is crucial for the image of the new public transport mode. A positive image is a powerful promoter for land and house price premiums.

e) At the start of the operation of a new public transport mode the vehicles have to work without any hitches otherwise a negative image is created which will take a long time to overcome.

f) According to the research carried out for EMTA (HASS-KLAU 2004) rail-based public transport modes are more powerful than buses in promoting economic development. In this respect the *light rail versus bus* argument is won by light rail.

g) Light rail lines and vehicles are more visible than underground railways and also offer views of the areas they pass through. It could be argued that both features increase or accelerate economic impact. New shops and restaurants can be seen from a tram/light rail vehicle but not from an underground railway. Hence although the number of passengers may be lower for light rail the economic effectiveness may well be higher.

Section five: Challenges in Dealing with the Complexity of Urban Transport

Introduction

In the final section of this anthology it is logical to focus on the challenges in dealing with the complexity of urban transport development. Some new ideas are presented in four chapters, one from North America, one from India and two from Europe.

In the first chapter of this section, the difficult role of political actors is analysed by Måns Lönnroth, Sweden, who has considerable experience of policy-making and politics. His starting-point is the observation that even carefully thought out plans for the change of urban transport systems are often delayed, changed or completely blocked by unforeseen and unforeseeable events. Such events are part of the everyday life of politicians.

To better understand how political decision-makers deal with the complexity of changing urban transport systems, more research is, according to Lönnroth, needed to analyse the relations between 1) the inner core (consisting of such actors as city mayors, the main political parties, the key city departments, public transport utilities, finance departments, property owners, construction companies, banks and so on), 2) the outer core (planners, architects, peripheral business interests, etc) and 3) the periphery of political decision-making (loosely defined actors or groups of citizens including transport researchers). One should also pay more attention to what kind of arguments that are regarded as 'important' in the political discussion.

Lönnroth here refers to the studies by the Danish sociologist Bent Flyvbjerg, who has analysed the relation between power and rationality. This researcher claims that those in power are able to defend their decisions by rationalising their motives. (He therefore introduces the /German/ concept of 'Realrationalität' as a parallel to the established concept of 'Realpolitik'.) Those outside the core of power are, however, obliged to refer to purely rational arguments if they want to influence political decisions.

The dilemma of politicians (in democracies) is to find a stable balance between what is considered efficient (which frequently concerns the views of those in the inner core) and what is considered legitimate (which concerns the views in the outer core as well as the periphery). This is perhaps the real meaning of the concept of "how to cope with the complexity of urban transport". In doing this, they may have to reflect upon whether the existing distribution of power between the inner core, the outer core and the periphery should in fact be redistributed. Current power relations might completely paralyse the political system and prevent it from being both efficient and legitimate in the social re-engineering of the urban transport system. Finally, Lönnroth emphasizes that much "could be gained from understanding the art of enforcing change and the extent to which this art actually differs between regions and between cities within regions".

In the second chapter of this section, David Marks from MIT begins by emphasising the key role of education and research in the search for solutions to the problems of urban transport. Against a background of current trends in urban transportation he claims that quite new initiatives are necessary. These trends are described in a number of figures illustrating 1) the increase of urban population all over the world, 2) the tendency of the population to contribute, by living in suburbs, to reduced urban density (in Western societies), 3) the emergence of new patterns of mobility – all this leading to 4) a heavy dependence on the automobile.

According to Marks, this development will challenge academia to work in an interdisciplinary way, and together with industry. Technologies will be the main tools, but their role requires understanding and linkage between technologies and social issues. Urban transportation systems are large scale complex engineered systems with multiple trade-offs: private vs. public, individual vehicles vs. public transport and passenger vs. freight.

Before going into detail about future research on urban transport, Marks enumerates four aspects of the entire transportation problem: 1) the dependence of transport on energy, particularly petroleum leading to unacceptable emissions of carbon dioxide, 2) emerging mobility issues (partly due to demographic changes), 3) the ongoing information and communication revolution and 4) fairly new security and terrorism concerns.

In the final part of his contribution, Marks describes what is going on within MIT to improve education and research on complex systems. His basic question is: How to provide adequate knowledge and how to implement it? Marks points to a number of tasks (both practical and theoretical) to be dealt with within urban transport development, for instance 1) providing accessibility for those without access to personal motor vehicles, 2) adapting the personal use vehicle to future accessibility needs/requirements, 3) finding ways to get the carbon out of transportation energy use as much/and as soon as possible, 4) improving the ability to model the technical and behavioural aspects of large-scale systems, 5) taking a step beyond the normal academic paradigm to real-world assistance in problem-solving.

While Marks' perspectives on urban transport development are Western (or rather Northern), and particularly American, Dinesh Mohan from Indian Institute of Technology Delhi develops a Southern perspective in the third chapter. Here, Mohan draws the reader's attention to the unacceptable lack of road safety in all countries. Traffic safety has certainly made progress in highly motorized countries. Less motorized countries are lagging behind in this development. Today's increasing use of cars and motorized two-wheelers add to these problems. This is a particularly serious fact in Asia and Africa, as urbanization is increasing in these parts of the world.

Mohan points to the fact that the emissions from the transport system (which is regarded as a general problem all over the world) will be substantially reduced only if more people are willing to walk, cycle and use public transport, but this will be possible only if walking and cycling (often a way to reach public transport) will become much safer in cities. Many cities are faced not only with a high level of road traffic accidents, but also with air pollution and crime on the streets, factors also reducing the willingness to walk and cycle.

Apart from the importance of politicians and planners, research has, according to Mohan, a role to play in addressing these problems. He can show that the researchers still do not know what factors influence road traffic fatality risks, pedestrian fatality risks and vehicle occupant fatality risks in cities with similar income groups and across different income groups. But he can show that cities with similar levels of wealth, knowledge and technology have very different road fatality risk ratios per capita.

Studies of factors influencing this situation should therefore be promoted, and – in many countries – more job openings should be created for those who want to choose a professional career in this area. He states that making cities safer for its citizens is also a prerequisite for providing cleaner environment.

In the final chapter of the anthology, Gunilla Jönson and Bengt Kasemo discuss how the Volvo Research and Educational Foundations should adapt its policy to the challenges of future urban transport development. They also invite other foundations and institutions, other researchers and experts, to take part in a learning process leading to a better understanding of problems and possibilities. The reason for this is the difficult task of breaking the path-dependency of the present development and of handling the tension between power and rationality. In the last section of their chapter they clarify their view of the means by which the aims of the Volvo Research and Educational Foundations can be realised in the near future.

Chapter 17. How do Political Actors Deal with the Complexity of Urban Transport Development?

Måns Lönnroth

Introduction

Someone once reputedly asked Harold Macmillan: "What is the most difficult thing about being prime minister? “Events, old boy, events” was the answer. The ever-flowing stream of events that disrupts order, the carefully thought out agenda of the day, foresight, planning, the much needed time for reflection and so on. Harold Wilson, another British Prime Minister, once said something similar: “A week is a long time in politics”.

A somewhat dismissive interpretation of these two quotations would be that the time horizon of politicians is at most the next election. Naturally, there is something in this interpretation – after all, politicians who are not re-elected can hardly be said to have been effective or successful. At the same time, the essence of politics – as in business – is split vision between the future and the present. Elections, likewise stock prices and consequently stock options, are always about the future. Any politician worth his or her salt must have a credible idea of the future. And to the extent that urban transport is a crucial element in the future of an urban region, urban transport will be a crucial element in the next election and the election after that and the election after that. There is nothing strange about this – my guess is that the same holds for CEOs (chief executive officers) in publicly owned companies. Strategies for future earning capacity should be as important as last quarter’s earning performance.

The comments of the two Harolds' point in another direction: political action and thus leadership is only partially amenable to rational planning. Unexpected events, or unexpected twists to expected events, can derail even the most logical and carefully thought out plan. Again, I would guess that this is not unique to politics. The history of war, and therefore of military strategy, is of course full of examples. Probably also the strategies of large companies.

At the same time, there are successes. Things do get done, in business, as well as in politics (and in some wars there are victories). In particular, urban transport systems do evolve, sometimes successfully and sometimes less so. The key question in my mind is: can we say something about the prerequisites for success in changing the future of urban transport? My own partial answer is yes: and the starting point should be to understand the role of power and how it is exercised in urban areas.

Power is as fundamental to the understanding of society as energy is to the understanding of nature. In English, the two words are often used synonymously. In both areas the key element is change. Energy drives change in nature and power drives change in society.

The Danish sociologist Bent Flyvbjerg is one of the most well-known Scandinavian social scientists. His books "Rationality and Power" and "Making social science matter" have been widely acclaimed as ground-breaking social science.

The first book is about traffic planning in Aalborg, a city that is the centre of a region of some 500,000 inhabitants. The author explains how traffic planning in Aalborg has evolved over a period of 15 years or so. To anyone with some experience in politics, the only comment is "at last". The story told has the obvious ring of truth. This is how things are, how they work out in real life. And it is about power, the role of rationality (by those not in power) and of rationalisation (by those in power). It is of course also about democracy and legitimacy and history and potential for redistribution of power and empowerment. In short, it is about the social engineering of the city transport system.

The theme of this book is about the complexity of urban transport. Complexity is then defined as a characteristic of a system; where the interactions between different parts are so many and so varied as to make understanding of the possible future paths of the system beyond rational analysis. More computer capacity is not enough, neither are more sophisticated models. There is a limit to human understanding – and this is particularly true of systems that include human beings as actors. There is no doubt that urban transport seen as a system has these characteristics, whether we are dealing with the American west or New Delhi.

So the answer to coping with the complexity of urban transport does not lie in engineering or in mathematical models or, for that matter, in economics. The notion of cost-effectiveness is at best a dead end – or at worst a decoy, a way to confuse the opposition. Every analysis of a complex system is bound to be a partial analysis. And every attempt to create a holistic perspective will require a comparison of apples, oranges, bananas and pears – both in terms of content and in terms of who stands to benefit from the conclusions. So where does that leave the academic, the politician and the people in between – those who propose a policy based on, hopefully, a sound analytical basis? I believe that we need to add a second definition of complexity, which is closer to how the word is used in everyday life.

Let us start with the politician. If an economist's first question to a proposed change is: "Is this cost-effective?", the politician's first question is: "Does it work?" I suggest that a successful politician, like any successful leader of an organisation in a complex environment, copes by finding the right angle of simplification by concentrating on what is really important as compared with what is simply important. But how is this done and what does this mean? This is probably the singularly most important area for academic research on the complexity of urban transport. Coping with complexity thus has two dimensions: the natural science dimension of understanding a system and the social science dimension of decision-making and leadership in complex environments.

In the social science dimension I believe that a deeper academic understanding of how politicians cope with changing the future of urban transport could cover three different areas:

Power relations

The first area: power relations. My guess is that achieving a better understanding of power relations must be done on three levels:

i) The inner core, which consists of all those actors that are always involved when important changes need to be made, be they small or large. These actors have long-standing relations with each other, sometimes adversarial, sometimes consensual but above all permanent. This is a group of actors that know that even if there is agreement – or disagreement – today there will always be a tomorrow with essentially the same actors. In this inner core, relations are both between individuals (which can go back a long way) and between institutions, which sometimes have a very long history – decades, even centuries. Members of this inner core include city mayors, the main political parties, the key city departments, such as roads and civil engineering, public transport utilities and finance departments, and also property owners (who are concerned about the impact on property values of changes in the transport system), property developers (with an interest similar to but not identical to that of property owners), construction companies (who stand to benefit from construction schemes), banks (who see financing opportunities but also risks), chambers of commerce (retail outlets) and so on.

ii) The outer core, which consists of all those actors are sometimes involved, depending on the issue at hand. Examples could be city environment departments, city land use planning officials, city architects, schools and day care centres (depending on local impact on traffic safety), more peripheral business interests, certain licensing authorities and so on.

iii) The periphery, which consists of all those loosely defined actors or groups of citizens who are only rarely asked to participate or who only by accident or sometimes sheer media savvy manage to get listened to. Neighbourhood NGOs (non-governmental organisations) and PTAs (parent-teacher associations) as well as transport researchers are typical examples.

One important, perhaps the defining, issue of the study of power relations is the tension between effectiveness – getting things done - and legitimacy – achieving the political acceptance of the outcomes, be they unintentional or intentional.

Getting things done is of course the essence of leadership. Leadership in urban regions is highly complex in itself, dependent on personalities as well as institutional structures. Both differ from time to time and from place to place. On the latter it is worth mentioning that metropolitan governance is a topic in itself worth studying. One important parameter is whether the region has one dominating centre or not; another is whether there is an overall metropolitan form of political authority or not. The role of national governments is also important. The Stockholm region of the 1950s and 1960s was a region with several independent cities (or local authorities). Nevertheless, strong leadership by the City of Stockholm managed to create a very efficient public transport system for the region as a whole. During later decades, the City of Gothenburg has been much more effective in mobilising political and economic capital to upgrade the regional transport system. In both cases, success (as well as a lack thereof) can be linked to specific individuals.

Legitimacy is the other part of the exercise of power. In one sense, lack of success in, for example, the Stockholm region to develop its transport system can be traced back to the lack of legitimacy for the steps proposed. But then legitimacy is a slippery subject. It is directly linked to presentation (by the centres of power) as well as perception (of those exposed to the outcomes). Here, the unintended but perceived outcomes are probably as important, if not more so, than the outcomes that are presented as the intended ones. This leads to the second area of study.

Arguments in public debate

The second area of study consists of understanding the role of arguments in public debate. Which arguments make a difference when someone inside or outside power wants to be listened to? Here, one has to make a distinction between arguments and analysis as presented in formal documents and arguments and presentation in the media and the public domain. A twist to a story in the media decides who is on the offensive and who is forced to defend. It decides who loses momentum and who gains momentum. Examples could be the role of travelling time for car-owners versus traffic safety for unprotected citizens, accessibility for goods transport, the type and time perspective of environmental improvements and deterioration, the cost of mobility for those who are already mobile, the cost of increasing mobility for those who lack mobility and so on.

Put somewhat crudely, one can make a distinction between rationality and rationalisation. Rationality is the only argument of the powerless; only through carefully constructed analysis can those outside the core of power hope to convince those with power. Conversely, rationalisation is frequently the argument of the powerful; it is much easier to get away with a spurious or downright false argument if you have power than if you do not.

Enter the role of academic research. In my view, the most important contribution made by academic studies of the complexity of urban transport involve the distribution of outcomes. Cheap and effective transport is important for everyone, but mostly to people with lower incomes. Who stands to gain from a proposed action and who stands to lose? This is, arguably, more important in the longer term than the idea of overall cost-benefit calculations based on discounted values. As we all know, political ideology plays a powerful role in deciding which arguments are considered valid by whom and thus considered as adding to legitimacy.

Social re-engineering or the redistribution of power

The third area of study concerns social re-engineering or the redistribution of power. Two questions immediately come to mind: redistribution within the inner core and redistribution between the periphery and the core. The second concerns inclusiveness, and this is linked to the development of analytical methods that can capture reasonably well how different, more or less marginalised groups are influenced by changes in the future of urban transport. The first includes the potential for autonomy of politics from powerful commercial interests. One example of redistribution of power within the core would come from congestion charges that are used to upgrade public transport (note that parking fees are also a form of congestion charge). Another example concerns the degree to which the cities – or those that finance the transportation system (both roads and public transport) can participate in capturing the increase in land value and property value that flow from increased accessibility. Since transport systems have a major impact on the future value of land, it is not far-fetched to discuss social re-engineering that makes it possible for the providers of transport services to capture some of the benefits of increased land value. A third example could include ownership as such of public transport – should it be centralised or fragmented? Obviously, studies of social re-engineering should also include financial instruments and the conditions under which experiences can as well as cannot be transferred from one metropolitan region to another.

In summary

I believe that one should not allow oneself to be too overwhelmed by the complexity of future urban transport. I doubt whether it is, inherently as a system, more complex than, say, the future of defence or the future of international finance markets. In fact, it may well be that there are lessons to be learnt from cross-sectorial studies of the management of complex systems. A great deal can also be learnt about the heterogeneity and the variations between different urban regions and much can be learnt about the inherent value (or crude-ness) of general recipes as proposed by international consultants or international financial institutions.

Much more could be gained from understanding the art of enforcing change and the extent to which this art actually differs between re-gions and between cities within regions. The US has a heated debate over smart growth, there are large differences between cities within Europe, and Australian cities appear to differ from Canadian cities and so on. African cities have characteristics that differ from Latin American cities. South Africa has the unique experience of land use planning driven by apartheid (and Israel is creating its own similar experiences on the occupied West bank). Indian cities differ from Chinese cities, reflecting differences in ethnicity, culture, history, political traditions and so forth. The structure of power, the role of argumentation and the prospects for redistribution of power all differ and offer rich areas for future academic research aimed at a better understanding of the potential for changing the future of urban transport.

My vision is that successful academic studies of the complexities of coping with the future of urban transport will result in tools and methods that can actually contribute to the empowerment of those with less access to power. The net result of such empowerment is of course a dilution of the power of the old core; in power relations there is hardly any room for win-win changes.

Chapter 18. Academic Studies of Complex Engineered Systems: How this New Knowledge can be Used to Improve Education and Practice for Mobility

David H. Marks

Introduction

A speech about academics and sustainability must begin with an examination of who we are and what we do. Among all the stakeholders of sustainable mobility, we are unique in our impartiality and lack of advocacy. While others advocate from their base and viewpoint, it is the role of academics to help illuminate dispassionately the salient facts and analyses and to educate others about them. We are the neutral platform upon which others can work to find synergies and agreeable solutions. Our job is to generate and accumulate knowledge and then transfer this knowledge to others who will use it. This is the model that has been in place for a thousand years well before books and blackboard; however, in a world being rapidly changed and stretched by information technologies, both new partnerships and different organizational structures will be needed to be effective. As such we must reform what we do, how we do it, and with whom we do it.

The role of research will be evolving and increasing in value. We do research to improve the education we provide and to help train future generations. Thus, any talk about education must also include a strong research component. The research component must be done not just by single researchers in isolation but also by groups with members of industry, government and the public. Similarly, more and more of what we do about research and education is driven by the problems of large-scale complex systems technologies; especially in that information technologies, materials, and energy sensors will be a main tool in our bag of tricks. By this we mean we must extend our knowledge and education of the methods of complex systems to insure a next generation of leaders of decision-making. This will involve changing our academic structure, which is a necessary but difficult task.

How we will use these technologies will relate heavily back to control methods - both physical and economic - and to the context in which they will be used. We need better linkage between the technical and the social. Each has been moving in parallel paths but we must now develop the insight, data, and models that will begin to interlink them. In the end we must understand behavioural pathways since they will provide the means to change the demand side of urban transportation

About Academia

Most of us come from strong disciplinary backgrounds and from a mode of work built around a silo of faculty members and associated research staff. Larger inter- disciplinary problems requiring cross-disciplinary teams and joint-work often are not valued, nor rewarded, in such an environment. This is a time when technically focused universities such as MIT are making a major change in their composition. Most major universities, in spite of tenure, have experienced a one-quarter to one-third turnover in faculty in the last five years. Further, these changes have not been in replacements but rather have been refocused into new directions such as information technologies, nanotechnologies, biotechnologies, neurotechnologies and macrotechnologies. It is in the last, particularly systems analysis and other macro techniques in aid of decision-making in complex systems that sustainable mobility may find a greater momentum towards good and equitable solutions.

Also of interest are structural changes in academia. MIT had not formed a new department in fifty years but in the late 90's it formed two new divisions within the School of Engineering. These new educational structures - the Engineering Systems Division and the Biological Engineering Division - can admit students, form curriculums for their fields and grant degrees. All involved faculty are two-key with each having a home in one of the Engineering School departments as well as within a Division. We also have new research structures: The interdisciplinary laboratory/centres (the Laboratory for Energy and the Environment which I head for example) help to bring more large-scale emphasis on cross-disciplinary areas. These, however, do not own faculty, do not admit students, and do not grant degrees but rather they help in pulling together cross-disciplinary teams and help to provide support for a great many of MIT's graduate students. All being driven by the fact that the most interesting problems are at the boundaries of the traditional disciplines.

What does this all mean for Sustainable Mobility?

Can these structures help to bring better focus from academia onto this area? There are many factors which make sustainable mobility a unique and highly complex systems problem. There are private investments as well as public investments in solutions with needed investment levels being very large. In the future we may not have the luxury of building new infrastructure but we will have to expand the capacity and quality of service of existing infrastructure with in the existing footprint.

We see vehicles being used for moving both people and freight, as well as for mass transit. Our road networks move both people and freight with the two functions often resulting in congestion. The system itself is the vital infrastructure for our economic growth and yet development is in danger of defeating itself through over-use, under investment, lack of proper managements and institutions, and little innovation in using price and technologies to smooth operations and increase capacities of networks.

These problems will not get easier. The world population will increase by about 50% in the next 30 years with almost all of these people living in the developing world. The developed world will increase very little but at the same time become more urbanized. This trend is greatly magnified in the developing world where the middle of the 21st Century will find almost 70% of the developing world living in urban settings rather than rural. People are moving from the poorer countryside to find work, education, security, food, mobility and other aspects made possible by density; but at the same time this density will make it more difficult to provide such services.

The problems are accentuated by the fact that urban population densities are falling in developed world urban areas. As people "suburbanize", they become even more dependent on individual vehicles and attendant infrastructure. The old, young, poor and disabled are denied mobility in such an arrangement and congestion and urban sprawl make mobility much more difficult to manage. The danger grows of congestion stifling economic activities and spreading environmental impacts.

There are important regularities in the patterns of the movement of people. As peoples' incomes grow, they travel more and they travel further. However, they do not spend more additional time or an increased share of their income to do so. Rather, they substitute faster means of travel for slower ones. "Faster" and "slower" relates to total trip time - not just to maximum vehicle speed. Access time, frequency of service, number of changes of vehicle required are all very important in determining mode's actual "speed".

All this helps to explain why public transport is losing market share to autos. The auto with its ability to allow sprawl and suburbanization has distributed people into places where a larger public system cannot economically be put into place.

The Future of Sustainable Mobility

The future shape, dimension, volume and value of sustainable mobility will be dependent on the extent of environmental impacts, the growth and control of urban congestion, and the impact on auto mobility, as well as emerging mobility issues. All of these are enumerated as follows.

1. Transportation is an enormous user of energy – especially petroleum

Transportation is overwhelmingly dependent on petroleum-based fuels. In 1999, transportation used 38 million barrels per day - just over half of world total of 75 million barrels per day. 96% of fuel used in transportation is petroleum-based (gasoline, diesel, residual fuel oil, jet fuel). Industrialized countries use two-thirds. And little, approximately about 43%, of transportation energy is used to move freight. Transportation's environmental dilemma is that it cannot avoid climate change issues; as virtually all transport systems produce carbon dioxide.

However there are significant differences in energy intensity across modes and across regions for the same modes, so that opportunities for efficiency increases will result in lower fossil fuel use per vehicle mile made possible in both the near- and long-term. Looking to future fuels it will be easier to move fixed (stationary) fossil fuel uses to carbon mitigation through carbon sequestration and the use of nuclear power and renewables than to move mobile uses. Thus mobility will be under the CO2 microscope.

However, transport may eventually cease to be a significant source of conventional emissions There has been progress in the developed world over the last quarter century in eliminating these emissions; however, this is not true in the developing world.

2. Emerging Mobility Issues

Changing demographics: Mobility for the elderly. With so little mass transit available, the old, young, poor, disabled and even those who do not want to use individual vehicles are not able to be mobile. Thus the places where they can live and the way they work and play are limited. New ways to give such people new access to mobility is essential. This can be through technologies, which will allow people to stay safely in cars for a longer period of time by enhancing guidance and vision technologies. This can be innovative management idea such as dial a ride, car sharing and rethinking of zoning and anti sprawl legislation. Thus we would expect to see vocational decisions changing to be more sensitive to congestion and location of where workers will live. We would expect to see trip making decisions made differently. Presently, trips to work make up only about one-third of total trips but nonetheless this has already changed. It is difficult to car pool to work if one needs to drop children at school on the way to work and do errands like shopping on the way home.

3. The information and communication revolution.

Can there be accessibility to knowledge without mobility? Many trips are to go to work, to gain knowledge and generally to participate in knowledge based activities. With the expansion of the internet and other information technologies, can we foresee a day when we work and play at home without the need to move about so much? The human character may not like this and there is evidence that improvements in information and communication may not slow the need for mobility. The introduction of the telephone was supposed to cut down on trips; however, the telephone expanded our number of contacts and allowed the family to spread out without losing contact. Both of these have engendered more need for mobility.

4. Security and terrorism concerns

Rethinking logistics and supply chain strategies became more apparent after the terrorism events of 9/11. Just in time systems proved to be very brittle with the breaking of just one link in the tree causing large manufacturing concerns to have to halt production. Already they are moving to diversify supplier, to keep more inventory at hand and to look for more flexibility in operation. All of this requires more mobility. The importance of mobility and mobility networks to economic performance is profound and there is a massive reexamination of how to make systems more secure and robust. Thus a vast system based on containers where it is hard to inspect even one percent is a large potential source for terrorism and disruption. More and more our highways and railways are used to transport hazardous products and wastes where even accidents or acts of nature - much less terrorism - can mean horrific damage.

Some observations as academics organize to help in these problems are:

- We are reaching the point where we have too much data - not too little

  But how is this data understood?

- We must bridge the large gap between the technical and its context
- How do we link the physical with the economics and the social?

Academia must evolve to produce a better graduate, more knowledge for problem solving across disciplines

How Will/Has MIT Organized to Provide Knowledge and Implement it?

1. Harnessing Existing Mobility Research Work at MIT

MIT has a long history of working in the automotive area with a variety of different foci. These include:

- Laboratory for Energy and the Environment
- Sloan Automotive Lab: Just completed a Wells-to-Wheels Power Train and *On the Road in 2020*
- 42 Volt Consortium for a new level of power source in autos
- Cooperative Mobility Program: Observatory on Mobility Practices around the World
- Safety Consortium
- Materials Selection: Lifecycle, Recycling Materials, Systems Lab
- International Motor Vehicle Program
- Program on Aging: Keeping People in Cars Longer
- Centre for Transportation Systems: Intelligent Vehicle Highway Systems

All of these work on different aspects of the Sustainable Mobility problem individually, cooperatively, and collaboratively with faculty and students in this area.

2. MIT Academic Challenge. Improve Education and Research on Complex Systems.

MIT is working to restructure its academic programs so that it can better cross-disciplines to bring about advances in complex systems understanding, methods, analysis, and problem solving. An example is the MIT Engineering Systems Division (ESD), whose role it is to understand, model, and predict the behavior of technologically enabled complex systems in order to address contemporary critical issues. ESD has 39 faculty and research staff devoted to teaching & research in engineering systems and who also have joint appointments in various Departments. ESD is an innovative crosscutting academic unit between most MIT School of Engineering departments, the Sloan School of Management, and the School of Humanities. It has approximately 300 students in five ESD Masters programs about 30 PhD students undertaking scholarship in systems, and has $20 million per year in research volume on engineering systems.

The mission and vision of the ESD:

Mission:

> Establish engineering systems as a field of study focusing on complex engineered systems and products viewed in a broad human, social and industrial context. Use the new knowledge gained to improve engineering education and practice.

Vision:

> ESD will be a leader in understanding, modelling, predicting and affecting the structure and behaviour of technologically enabled complex systems.

ESD will focus on six application domains of expertise around three intellectual foci in conjunction with the MIT School of Engineering Departments and MIT School of Management. ESD has five proposed integrative thrusts.

ESD Faculty work at many different levels of complex systems:

- Societal level
- Enterprise level
- Artifact level

They learn about what is common and essential to systems in the abstract by constructive comparison of various systems (air, manufacturing, etc)

Understanding Engineering Systems requires:

- Interdisciplinary Perspective – technology, management science and
- Social science
- Incorporation of system properties such as sustainability, safety and
- Flexibility in the design process.
- Enterprise Perspective
- Different Stakeholder Perspectives

ESD Seminal Questions

- Can system architecture be moved up the hierarchy of knowledge from description to prediction?
- Can enterprises be predicatively architected?
- Can system properties such as safety and sustainability be approvingly designed into systems?
- Can socio-technical interactions be predicatively modelled?
- Can flexibility be analytically built into system design?

Academic Challenges in Urban Transportation

1. Methods to provide accessibility for those not having access to personal motor vehicles- redefine the relationship between public transportation, the private car and bringing knowledge and interaction to the home.
2. Adapt the personal use vehicle to future accessibility needs/requirements (capacity, performance, emission, fuel use, safety, materials requirements, waste, ownership structure, etc.)
3. Looking for ways to get the carbon out of transportation energy use as much/and as soon as possible.
4. Serve as a convener of the debate about the future directions of sustainable mobility. In a world of split responsibilities and separate technology development, academia has a good perspective for pulling all these disperse pieces together. This can be done by serving as the platform for the various stakeholders to meet, looking for innovative ways to solve problems, capturing best practices around the world, and evaluating new ideas from the stakeholders to give them independent credibility.
5. Building the new academic component of large-scale complex management using ESD and other evolving programs around the world to build new multidisciplinary structures for research and educational programs
6. Improving our ability to model the technical and behavioral aspects of large-scale systems. Combine new advances in computation such as systems dynamics, with our increased ability to capture real time data and with better partnerships with social scientists to better understand the technology/society interface.
7. Taking a step beyond the normal academic paradigm to real-world assistance in problem solving.
8. Methods to provide accessibility for those not having access to personal motor vehicles. Redefine the relationship between public transportation, the private car, and bringing knowledge and interaction to the home.
9. Adapt the personal use vehicle to future accessibility needs/requirements capacity (performance, emission, fuel use, safety, materials requirements, waste, ownership structure, etc.)

10. Continue the focus on more efficient deliveries in urban areas using a systematic all-encompassing approach, which would include:

    - Simulation of the impact of a growing fleet of urban delivery vehicles in traffic conditions
    - Analysis of the energy and environmental impacts of existing and new freight distribution concepts
    - Seeking consolidation of services and operations will be of great importance but which make the most sense?
    - The impact of future vehicle and information technology on freight transportation and the environment in which it operates
    - The optimum balanced mix of policy options to reduce environmental impact of urban goods distribution and the environment it operates in while making operations more effective

In conclusion, the recognition by academia that complex systems problems are of great importance will require new ways of operation. Interdisciplinary work bringing together technology, management and the social sciences will be needed. New partnerships with industry government and the public will be required. Together we will learn novel ways to solve a problem that is at the heart of our future sustainability.

References: Most of the figures and statistics are from an MIT Report *Sustainable Mobility 2001* done for the World Business Council on Sustainable Development. Copies may be downloaded from www.wbcsd.org. Since this talk was given in September 2003, a July 2004 report *Sustainable Mobility 2030* has been issued by WBCSD and is available as well.

Information on the MIT ESD can be found at esd.mit.edu.

Chapter 19. Safety and Sustainable Urban Transport

Dinesh Mohan

Introduction

Most of the large cities in the world are already located in less motorised countries (LMC) and many more cities in these countries are expected to have populations of ten million or more in the next few decades (1). All these cities are faced with serious problems of inadequate mobility and access, vehicular pollution and road traffic crashes and crime on their streets. Increasing use of cars and motorised two-wheelers add to these problems and this trend does not seem to be abating anywhere. However, many recent reports suggest that improvements in public transport and the promotion of non-motorised modes of transport can help substantially in alleviating some of these problems (2-5). The problem is that current evidence shows no success in reducing the use of personal motorised transport for long distance trips anywhere in the world and for urban trips in most locations (6-8). Most efforts to reduce environmental pollution due to road transport, therefore, focus on the control of exhaust emissions. This has produced some successes in reducing CO, SO_2 and NO_X in a few locations, but not CO_2 anywhere. As long as we use fossil fuels for combustion this problem is unlikely to be resolved unless we can shift modal shares towards non-motorised and public transport. In this paper we discuss the essential role of traffic safety in promoting more sustainable modes of transport in both LMC and highly motorised countries (HMC).

LMC cities have very mixed land use patterns and a very large proportion of all trips are on foot or by bicycle; of the motorised trips more than 50% are by public transport or shared para-transit modes.

Compared to HMC, trips per capita per day in LMC are lower and a significant proportion of trips can be less than 5 km in length; and the cost of motorised travel is high compared with average incomes (3). In spite of these structural advantages, the air pollution levels in LMC cities remain high.

In addition to the problems of pollution, deaths and injuries due to road traffic crashes are also a serious problem in LMCs (9). Even in HMC pedestrians and bicyclists generally face higher crash risks than car occupants (10-12). According to one estimate the losses due to accidents in LMCs may be comparable to those due to pollution (13). These problems become difficult to deal with because there are situations in which there are conflicts between safety strategies and those that aim to reduce pollution (14). For example, smaller and lighter vehicles can be more hazardous but they are less energy-consuming; congestion reduces the probability of serious injury due to crashes but increases pollution; an increase in cycling rates can decrease pollution but may increase collisions if appropriate facilities are not provided. Integrating safety and clean air policies is not an easy process as the professionals dealing with the two problems seldom talk to each other. Furthermore, both issues are reasonably complex as they demand the interplay of scientific, technical, social and political concerns. Historically, societies have found it difficult to put in place integrated safety policies as many interventions based on "common sense" have not proved to be successful.

Ethical base for sustainable transport

A stencilled outline of a body memorialises the spot on 74th Street and Central Park West in New York City where Henry Bliss was run over by a cab on September 13, 1899, becoming the first known pedestrian fatality in North America (15). In the same year UK recorded its first motor car driver fatality in London (16). Internationally, there are now more than a million fatalities annually, and the number is still on the rise (17). The concern with road safety has very early origins but lack of professionalism still pervades the field of traffic safety. Trinca et al make the following comment in the award-winning book *Reducing Traffic Injury –A Global Challenge* published more than fifteen years ago (18)

> Anyone can don the mantle of 'expert' and many citizens are convinced they have the 'answer'. Articulate lobbyists, not always with proper qualifications but often with the best intentions, press for measures which frequently are not well founded.

This is partly because road safety has still not developed as a scientific course of study in most countries and there are very few job openings for those who want to pursue a career as a full-time road safety professional. Because of the widespread concern with road traffic deaths in the middle of the twentieth century in North America and Western Europe, there have been major advances in vehicle safety technology, road design, traffic law enforcement techniques, and post-injury management. But, policy-makers and road safety professionals in every country have found it very difficult to institute changes that actually result in a dramatic decrease in traffic fatalities and injuries in a short time. International experience has also shown that not all individuals follow all the instructions given to them to promote road safety. Attempts to "educate" people regarding safety have also not always been very effective in any country and wide variations are found between people's knowledge and their actual behaviour (19).

It is in this situation that Vision Zero was included in the Road Traffic Safety Bill passed by a large majority in the Swedish Parliament on 9 October 1997. Vision Zero may be summarised as follows (20;21):

- The scientific basis of Vision Zero differs from the usual approach to safety in human-machine systems: designing a system to minimise the number of events that cause injury. Instead, Vision zero is based on the notion of "allowing" these incidents to occur, but at a level of violence that does not threaten life or long-term health

- In Vision Zero, the entire transport system must be designed to accommodate the individual who has the worst protection and the lowest tolerance of violence. No event must be allowed to generate a level of violence that is so high that it represents an unacceptable loss of health for that vulnerable individual
- The responsibility for every death or loss of health in the road transport system rests with the person responsible for the design of that system. This is the ethical basis for realising Vision Zero

The ethical basis is underscored by the adoption of the Montreal Declaration on People's Right to Safety (22) in environments where the use of products and services is not necessarily the choice of individuals. This is true for transportation as road users cannot opt for not using the system. We therefore have a societal and moral responsibility to design our products, environment and laws so that people find it easy and convenient to behave in a safe manner without sacrificing their need to earn a living and fulfil their other societal obligations. The systems must be such that they are safe not only for "normal" people but also for those individuals who may be disadvantaged socially, economically or in their psychological or motor abilities. These kinds of designs, rules and regulations would reduce the probability of people hurting each other or themselves even when they make mistakes. Such changes will take place in a systematic manner only when safety is recognised as a fundamental right of communities and does not depend just on the goodwill of powerful institutions (23). In this paper we show that the establishment of this ethic is a necessary condition for establishing sustainable transport policies in the future.

Issues influencing sustainable transport policies

The majority of people in all large cities around the world use motorised modes of transport for their work trips. This proportion is likely to increase as urbanisation increases in Asia and Africa. In Europe, road transport now makes up 44% of goods transport and 79% of the total passenger transport market (5). In the U.S. the car is most dominant for the work trip, accounting for 97% of all journeys to work in rural areas and 92% in urban areas (7). In LMC large cities use of motorised transport is significant and is increasing (8;24). For sustainable transport policies we need to arrest the increasing trend of the adverse health and environmental effects of road transport. The following issues need to be addressed:

- Motor vehicles have killed more than 20-30 million people and injured more than 500 million worldwide in the last one hundred years. This is not sustainable
- Emissions will reduce significantly only if more people walk, cycle and use public transport. This will be possible only if walking and cycling is made much safer in our cities
- Cities will only improve on an aesthetic, humane and human scale if streets include large numbers of people walking and playing safely

Since walking and cycling are essential for use of public transport, streets must be made safe from crime, disabled friendly and include public amenities, such as shops and restaurants (street vendors included). These conditions can only be fulfilled if special attention is given to speed-reducing measures along with street designs fulfilling traffic-calming designs. Figure 19.1 shows the road traffic fatality risk per million persons in different cities plotted against per capita income of the country in which they are located. These data show that the risk varies by a factor of 20 between the cities with the lowest risk and the highest risk. Detailed analyses for factors associated with crashes for different cities are not available. However, some general observations can be made on these issues.

- The highest fatality rates seem to be experienced by cities in the mid-income range of USD 2,000 – 10,000 per person per year
- Overall fatality risk in cities with very low per capita incomes (less than USD 1,000) and those with high incomes (greater than USD 10,000) seem to be similar
- There is a great deal of variation even in those cities where the per capita income is greater than USD 20,000 per year

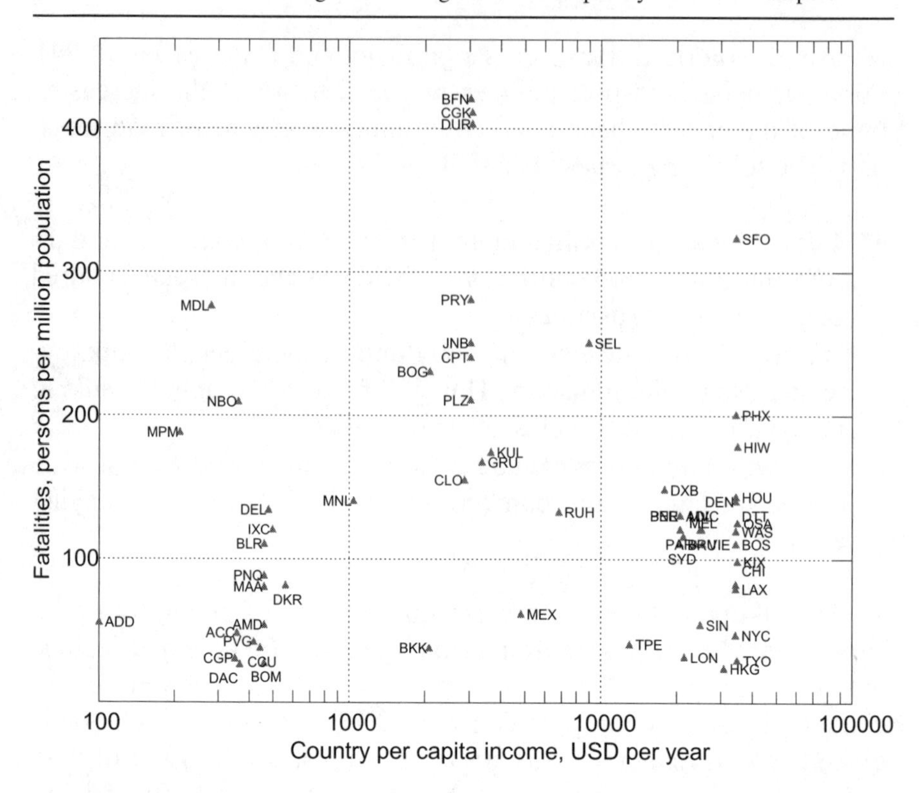

Figure 19. 1. Road traffic fatality risk per million persons in different cities by per capita income in US dollars.

These patterns appear to indicate that it is not enough to have the safest vehicle technology in high-income countries to ensure low road traffic fatality rates uniformly across cities in those locations. Even in very low-income countries, the absence of funds and possibly unsafe roads and vehicles does not mean that all cities have high overall fatality rates. We get some idea of why this may be happening if we examine vehicle occupant (excluding motorcycles) fatality rates and pedestrian fatality rates per million populations of the cities as shown in Figures 19.2 and 19. 3.

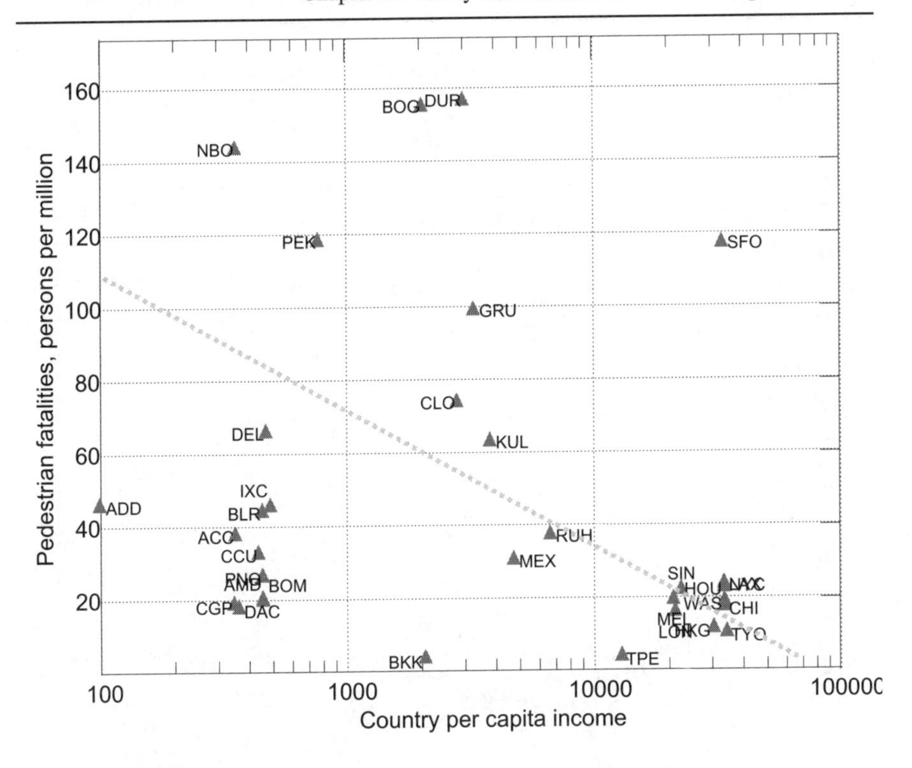

Figure 19.2. Pedestrian fatality risk per million population for different cities by annual per capita income.

These figures show that pedestrian fatality rates are consistently low in high-income cities (greater than USD 20,000 per capita per year), but can be high in low-income and middle-income cities (USD 300 – USD 10,000). On the other hand vehicle occupant fatalities rates are generally higher in high income cities compared with low-income cities. If we consider the patterns in Figures 19.1, 19.2 and 19. 3 jointly it is possible to make some preliminary observations on these issues.

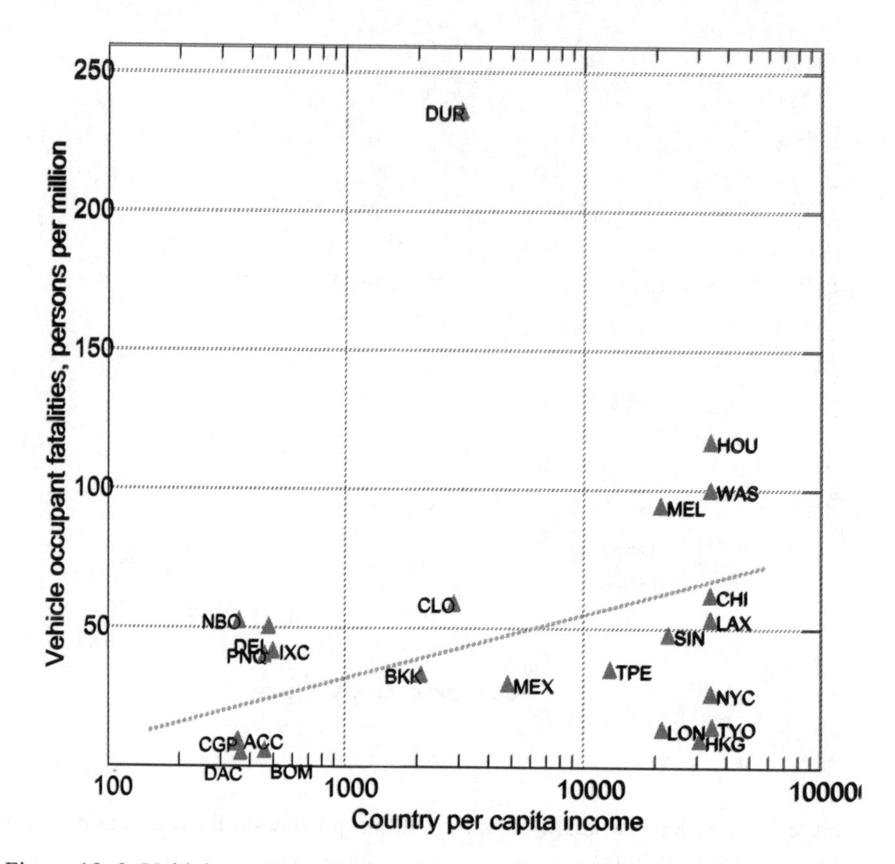

Figure 19. 3. Vehicle occupant fatality risk per million population for different cities by annual per capita income.

1. Provision of safely designed roads and modern safe vehicles may be a necessary condition for low road fatality rates in cities but not a sufficient one.

The fact that there are wide variations in overall fatality rates among high income cities, where availability of funds, expertise and technologies are similar, indicates that other factors, such as land use patterns and exposure (distance travelled per day, presence of pedestrians, etc.) also play a very important role. This is probably why many European cities tend to have lower rates than those in the US.

2. Studies published over the past few years show conclusively that vehicle speed is very strongly related to both the probability of a crash and the severity of injury – a 1% increase in the average speed can result in a 3-4% increase in fatalities (9). This may be the reason why some middle-income country cities have high fatality rates because they have higher vehicle ownership than low-income countries and roads encouraging unsafe speeds without adequate attention being given to road safety. Similarly, cities that are considered to have greater traffic congestion (hence lower speeds) have lower rates than those with less congestion though their incomes may be similar – Mumbai (BOM) in India with higher congestion has lower rates than Delhi (DEL) in India with lower congestion levels, and New York (NYC) in the US has a lower rate than Houston (HOU), also in the US.

3. Figure 19.2 shows that cities in high-income countries (> USD 20,000 per capita income) have much lower pedestrian fatality rates than most cities in low- and middle-income countries. This could be because high-income countries may have a much lower presence of pedestrians (25) than low-income countries and also because there has been a conscious effort in the former to increase pedestrian safety.

4. Figure 19.3 shows that, in general, cities in low-income countries have lower vehicle occupant fatality rates (per million population) than cities in high-income countries. This is probably because vehicle ownership rates are much lower in low-income countries than in high-income countries, resulting in much lower exposure rates in the former. What is of interest is that the vehicle occupant fatality rates within high-income cities differ by more than a factor of 5. This is

clear evidence that just the presence of "safe" modern vehicles is not an adequate condition to minimise the health burden of road traffic injuries.

The above discussion indicates that safety of road users in general and pedestrians in particular has not been maximised given the present state of knowledge. Land use policies that encourage greater use of cars per day and transportation policies that promote personal transport increase exposure rates and thus the overall risk of death and injury even though risk rates per km travelled may be low. This tends to offset the advantages gained by the provision of safer cars and roads. Once these systems are in place, it is difficult to reduce death rates per capita.

These issues must be examined in greater detail so that we can plan for safer cities in the future in addition to improving vehicle and infrastructure technologies. This would be essential for cities in low- and middle-income countries where urbanisation is still increasing. For sustainable transport policies it would be essential that these cities do not get locked into systems that encourage high speeds and greater use of personal car transport per day.

The way forward

Road safety promotion as a sustainability issue

Road safety in general and safety of vulnerable road users in particular must be assigned the same importance as vehicle emissions for ensuring cleaner and more liveable cities. Unless cities are made safer for pedestrians and cyclists on the one hand, and women, children and the elderly on the other, it would be impossible to obtain optimal use of public transport facilities in the future. Road safety must therefore be included as a necessary condition for healthier life in cities.

Understanding urban settlements and transportation systems

The discussion in previous sections illustrates that we do not know enough about different factors influencing the health burden of road traffic injuries and deaths on society. This is shown by the fact that cities with similar levels of wealth, knowledge and technologies have very different road fatality risk ratios per capita. It is essential that we promote international comparative studies of cities within similar income groups and across different income groups to understand the influence of factors other than vehicle technology. Some of the issues that can be examined are:

- The location of low-income neighbourhoods and their influence on travel patterns and road traffic injuries
- The presence of street vendors and their influence on vulnerable road users and street crime rates
- The influence of high-rise versus low-rise habitation
- The presence of a few large central business districts versus several smaller, widely distributed business districts
- The influence of very high-capacity public transport systems (underground or elevated metro systems) versus high-capacity modern bus transport systems
- The role of dedicated bicycle lanes and the use of non-motorised transport
- The role of para-transit and micro-vehicles, such as three-wheeled scooter taxis, etc

References

(1) World Health Organisation. Healthy cities: Air management information system. AMIS 2.0 CD. 1998. Geneva, WHO.
(2) Wu Yong, Li Xiaojiang. Targeting Sustainable Development for Urban Transport. 1999. Beijing, CICED. Workshop on Urban Transport and Environment.
(3) Mohan D, Tiwari G., Saraf R, Kale S, Deshmukh SG, Wadhwa S et al. Delhi on the move: Future traffic management scenarios. 1996. Delhi, Transportation Research and Injury Prevention Programme, Indian Institute of Technology.
(4) Synthesis report on environmentally sustainable transport: Futures, strategies and best practices. Vienna: Austrian Ministry of Agriculture, Forestry, Environment and Water Management, 2000.

(5) Commission of the European Communities. White paper. European transport policy for 2010: Time to decide. Com (2001) 370 Final. 2001. Brussels, Commission of the European Communities.
(6) European Commission. European transport policy for 2010:time to decide. 1-118. 2001. Luxembourg, Office for Official Publications of the European Communities. White Paper.
(7) U.S. Department of Transportation. 2001 National Household Travel Survey. Washington, DC: Federal Highway Administration and Bureau of Transportation Statistics, 2003.
(8) World Business Council for Sustainable Development. Mobility 2001: World mobility at the end of the twentieth century and its sustainability. Geneva: WBCSD, c/o E&Y Direct, 2001.
(9) World report on road traffic injury prevention. Peden Margie, Scurfield R, Sleet D, Mohan D, Hyder AA, Jarawan E et al., editors. 2004. Geneva, World Health Organization.
(10) Pucher J, Dijkstra L. Making walking and cycling safer: Lessons from Europe. Transportation Quarterly 2000; 54(3):25-50.
(11) Jorgenesen NO. The Risk of Injury and Accident by Different Travel Modes. International Conference on Passenger Safety in European Public Transport. Brussels: European Transport Safety Council, 1996: 17-25.
(12) Scientific Expert Group on the Safety of Vulnerable Road Users (RS7). Safety of vulnerable road users.
DSTI/DOT/RTR/RS7(98)1/final, 1-229. 1998. Paris, Organisation for Economic Co-operation and Development.
(13) Vasconcellos EA. Urban Development and Traffic Accidents in Brazil. Accident Analysis and Prevention 1999; 31(4):319-328.
(14) OECD. Integrated Strategies for Safety and Environment. 1997. Paris, Organisation for Economic Co-operation and Development.
(15) Perils for Pedestrians - Episode 41.
http://www.pedestrians.org/episodes/details31to60/episode41.htm . 2000. 19-04-2003.
(16) The history of road safety, traffic legislation and other important motoring events in UK.
http://www.roadsafetyuk.co.uk/histall.htm . 2003. Road Safety UK. 17-04-2003.

(17) Jacobs G, Aeron-Thomas A, Astrop A. Estimating Global Road Fatalities. TRL Report 445. 2000. Crowthorne, U.K., Transport Research Laboratory.
(18) Trinca GW, Johnston IR, Campbell BJ, Haight FA, Knight PR, MacKay GM et al. Reducing Traffic Injury - A Global Challenge. Melbourne: Royal Australasian College of Surgeons, 1988.
(19) Robertson LS. Injuries: Causes, Control Strategies and Public Policy. Lexington, MA: Lexington Books, 1983.
(20) Tingvall C. The Zero Vision. A road transport system free from serious health losses. In: Holst von H, Nygren Å, Thord R, editors. Transportation, Traffic Safety and Health. Berlin: Springer Verlag, 1997: 37-57.
(21) Elvik R. Can injury prevention efforts go too far?: Reflections on some possible implications of Vision Zero for road accident fatalities. Accident Analysis & Prevention 1999; 31(3):265-286.
(22) Montreal Declaration on People's Right to Safety. http://www.iitd.ac.in/tripp/righttosafety/Montreal%20declaration%2015-05-02.htm . 2002.
(23) Mohan D. Safety as a human right. Health and Human Rights 2003; 6(2):161-167.
(24) Tiwari G. Towards A Sustainable Urban Transport System: Planning For Non-Motorized Vehicles in Cities. Transport and Communications Bulletin for Asia and the Pacific 1999; 68:49-66.
(25) Select Committee on Environment and Regional Affairs. Walking in Towns and Cities. Eleventh Report. 2001. House of Commons, London.

Chapter 20: How does VREF Proceed?

Gunilla Jönson and Bengt Kasemo

This anthology is part of the programme initiated by Volvo Research and Educational Foundations (VREF) a few years ago to support new and equitable solutions for urban transport. As Bengt Kasemo has said in the introduction, the conferences in 2000 and 2003 have shown that a holistic view and a systems approach appears highly desirable, or even necessary, when problems and solutions have to be addressed in this area. This view actually derives from Tengström's statement that it is a matter of "copying with complexity".

The development of urban transport systems today seems to be subject to *market failures* as well as to *government failures* and *interaction failures* as emphasised throughout the anthology. Present knowledge indicates also that it might be useful to add the concept of '*public acceptance failure*', referring to the resistance of the general public to accept various proposals to reduce the problems of an urban transport system.

We are all very much in a learning process. The area is 'complex' in the sense discussed by both Tengström and Kasemo at the outset. We need to understand this complexity and the various interactions of the components and actors in this system, how and when planning and implementation is possible, to what extent solutions can be translated from one urban area to another, the relative importance of hard components (technology etc) and soft components (interactions of actors etc), and so on. This colours the planning and future efforts of VREF. Analytical work, empirical data collection, simulations and modelling, and implementation dedicated studies/actions are all necessary. It is a challenge to find a good balance between the activities in these sub-areas, and to make them interact in a cross-disciplinary way, rather than developing into non-interacting disciplines.

VREF intends to support continued studies and discussions in the FUT area through research grants as well as conferences on the complex socio-technical and economic systems that constitute urban transport. This work will be successful only if the Foundations manage to receive the necessary continuous and increased support from actors, institutions (both material and immaterial) and organisations throughout society. It is involvement by individuals as well as organisations that is necessary if we are to succeed. No single body will be able to handle all the problems and solutions. Therefore, co-operation across organisation boundaries and interest areas is important. This might also include co-operation with other funding agencies. Hierarchical structures have to be broken down and new ways of cooperation have to be found. VREF is not a major player in the arena. The hope is, nevertheless to make a difference by being a catalyst and lubricant in building an activity, where learning and actions go hand in hand, and to contribute to new or improved solutions in the FUT arena.

Several of the contributors to this anthology stress rapid urbanization in the near future. This growth is taking place all over the world. In the developing world it has often to do with an ongoing unplanned immigration of poor people to the cities. Their economic activities are mostly informal, but the growth of the formal economy is still partly involved. All these activities require movements that have to be facilitated in different ways.

The future supply of energy is a critical factor for society as a whole, including its transport systems as a major consumer of energy. It creates uncertainty and quantitatively unknown boundary conditions for the future. We know that the fossil fuels are declining, but not how fast, and with a quantitatively unknown demand and price development implying unknown future costs of transportation. To what extent can we prepare ourselves for this and associated factors? As Kasemo says in his introduction, we might *hope* and push for new (sustainable) energy supply chains developing faster than we see today, but we cannot build the next few decades on such beliefs. Instead, we have to take into account, and be prepared for the built-in uncertainty in the energy arena. This is indeed a major challenge. As if this were not enough, this issue is intimately linked to another major challenge for society: to build and maintain a sound environment, locally and globally, which in turn is coupled both to health issues and to the general life conditions on earth.

Energy and environmental challenges belong to the overall challenge of building what we today call a sustainable society. Energy supply is one of the cornerstones, but it includes much more. All actors have a responsibility, as we see it, to aid in the building of a sustainable society, including its production of goods and services, housing, water and food, a good environment, work for people, an equitable distribution of resources, etc.

There is no question that a society meeting human demands and needs requires a sustainable urban transport system. The detailed development and design of Future Urban Transport (FUT) systems will be affected by the energy and other boundary conditions. The existing cities today, and their transport systems, have developed along *paths* that we can analyze with regard to earlier boundary conditions, power and decision structures, historical, cultural and societal boundary conditions etc. To study the paths of the past, and combine them with the identifiable future boundary conditions, like the energy challenge, is hopefully a way of better distinguishing potential paths for the future. In this context the question already mentioned of "what is specific and what is generic" is highly relevant. It is obviously important to understand what is general and what is specific for the different paths of development. If we can distinguish specifics from generics, it increases the chance that we can assess transferability and non-transferability. A smooth and evolutionary transition of today's systems into the systems of tomorrow may be possible, and is, of course, what we strive and hope for, but a more disruptive scenario is not unlikely. We just don't know, but through understanding and planning we can increase the probability for the sustainable path. This is part of what we call *path dependencies* and of "*coping with complexity*".

Intensified studies and discussions about the complex processes of change are going on, and ought to go on to an even greater extent. The special emphasis should then be on what driving forces there are for change, how different actors in the inner and outer core and in the periphery, as identified and discussed by Måns Lönnroth, can influence the process. The studies and discussions should also pay attention to what the relation is between power and rationality as well as that between power and efficiency, effectiveness and legitimacy. The driving forces need to be identified in political changes, as well as changes depending on the world energy situation and brutal environmental impacts on living conditions for great numbers of people. The processes may seem turbulent at present, but there is no major challenge to the prediction that there will be trends and breakthroughs that are even more brutal. It will influence prices on energy to keep societies moving, and it will have an impact on environmental restrictions through new agreements of the Kyoto type

The above and the introductory chapters provide some of the main perspectives for VREF's planning and the work ahead (see also the policy document for VREF's work, www.volvoresearchfoundations.com). One important component of this work is the conviction that it is important to involve both analysts and actors in the future work.

Finally we provide a few brief comments for the future about the three "instruments" (Centres of Excellence, Smaller Projects and the FUT conference) with which VREF works.

1. Besides the three Centres of Excellence already established in India and the USA, we aim to establish new Centres of Excellence in Europe, Latin America, Africa, Asia and/or Australia, at a rate of one centre each year or every other year, in order to ensure a build-up of knowledge of the urban transport systems in different parts of the world. Planning grants have already been granted for proposals from some of these continents. VREF hopes and believes that these appli-

cations will develop into fully-fledged long-term Centres of Excellence.

These existing and new Centres are expected together to create a global network, making it possible to share knowledge through research educational collaborations, exchange of research results and ideas, implementation schemes and comparisons of experiences. The network may be built both through the Internet and through the exchange of young and senior researchers; the form is not so important. The important goal is to establish a network of centres that can mutually interact and learn from each other, in order ultimately to contribute to the long-term building of sustainable transport systems.

2. The Centres of Excellence are complemented with Smaller projects that may deepen the knowledge either in certain subareas of the CoE activities or in areas that lie outside but complement the CoE activities. One such example is research that complements and deepens our understanding of what is generic and what is specific. A second example is relationships between different actors in the inner and outer core and in the periphery, and their influence on the process of developing FUT systems. We find it especially important to develop understanding and knowledge about Asian, Latin American and African cities.

3. The Future Urban Transport Conferences are aimed at becoming a natural meeting place for analysts and actors who participate in the building of knowledge and understanding in the FUT area. As the number of CoEs and smaller projects increase, participation in them is likely to increase. However, we emphasize that the goal is a general meeting place, independent of whether participants are active in VREF projects or not. The real goal is to create a meeting-place with increased interaction between politicians, planners, industrialists and the grant-holding researchers. We believe there is a growing need for all these individuals to interact and un-

dersstand each other. The conferences in 2000 and 2003 had as their the main theme how to cope with the complexity of the urban transport development. The conference in 2006 will discuss how we have learnt to cope with complexity so far, and what we can learn from this for the future. The reader of this anthology may follow the development of the planning of the conference in 2006 at www.fut.se. Contributions and viewpoints are appreciated.

List of Figures, Graphs, Tables and Boxes

Some useful addresses

www.volvoresearchfoundations.com

This website presents information on the Volvo Research and Educational Foundations (abbreviated VREF). It contains information on the current programme on Future Urban Transport (FUT), the FUT Policy Statement document, instructions for how (and when) to apply for research funding within the FUT programme, and full contact information. You will also find all useful addresses (phone, fax and e-mail).

www.fut.se

This website provides information on the FUT conferences organised by VREF in Göteborg, Sweden. It covers both the two previous conferences, held in 2000 and 2003, and the upcoming conference planned for April 2006.